Human Genetics

SECOND EDITION

Human Genetics

AN INTRODUCTION TO THE PRINCIPLES OF HEREDITY

SAM SINGER

University of California, Santa Cruz

W. H. Freeman and Company
New York

Library of Congress Cataloging in Publication Data

Singer, Sam, 1944–
   Human genetics.

   Includes bibliographies and index.
   1. Human genetics.   I.   Title.
QH431.S638   1985   573.2′1   84-28632
ISBN 0-7167-1648-8 (pbk.)

Printed in the United States of America

1 2 3 4 5 6 7 8 9 0   HL   3 2 1 0 8 9 8 7 6 5

Contents

Preface

Most people are naturally interested in the subject of genetics, especially as it relates to the inheritance of physical and behavioral traits that run in human families. Nonetheless, genetics has the reputation of being a difficult subject, and many people deprive themselves of the fun of understanding how genetics works because they wrongly assume that the subject is too complicated for them. I have written this book in the hope of sharing with you what I think are the fundamentals, and some of the interesting highlights, of human genetics. In the seven years since the publication of the first edition of this book, the field of genetics has exploded once again. The second edition reflects this mushrooming of knowledge in that it is longer and somewhat more detailed than its predecessor. But the goal of the narrative remains the same. Anyone with enough interest in the subject to pick up this book and browse through it will probably be able to understand what is written here, regardless of previous background in biological science.

The book has eight chapters. The first two explain the inheritance of some characteristics that run in families in terms of chromosomes and account for the fact that the human population is made up of nearly equal numbers of males and females. Chapter 3 explains how the material of heredity, DNA, produces its effects and how the structure and function of proteins depend on their precise three-dimensional architecture. Chapter 4 deals with the regulation of gene expression and with the genetic basis of certain kinds of human cancers. Chapter 5 discusses the categories of DNA that make up the human genetic program and explains how maps of human chromosomes are constructed. Chapter 6 is concerned with the prenatal diagnosis of genetically determined abnormalities, an area in which much progress has been made in the last few years. Chapter 7 is about the genetics of human populations. Here the emphasis is on genetically determined differences in protein molecules and on how heritable changes in the genetic program, or mutations, ultimately arise from accidents that affect DNA molecules. Finally, Chapter 8 explains how some aspects of human behavior depend on the interactions of genes and the environment, and the book concludes with some speculations about human evolution in the future.

For those who enjoy such things, the text is followed by an appendix that contains some problems pertaining to the patterns of inheritance discussed in the first two chapters. This is followed by another appendix on basic chemical principles and structural formulas.

I sincerely thank once again the following professors who read, commented on, or otherwise helped to improve my rough drafts for the first edition: Cedric I. Davern, Robert S. Edgar, Ursula W. Goodenough, and Henry R. Hilgard. The second edition greatly benefited from the additional comments of professors Henry R. Hilgard, R. L. Bernstein, and Y. Edward Hsia.

Thanks are also due to John Staples and to my friends at W. H. Freeman and Company, especially Linda Chaput, and to Jim Dodd and Heather Wiley. And finally, thanks to everyone who reads this book, enjoys it, and learns something, as I surely did while writing it.

Sam Singer
Felton, California
March 1985

Human Genetics

Chapter 1

Traits and Chromosomes

In the mid-eighteenth century the city of Paris was the scene of a most unusual mating that aroused widespread public interest. The affair concerned a male rabbit who unexplainably showed great sexual interest in a certain barnyard hen. The hen, for her part in the matter, readily tolerated the rabbit's advances but would have nothing to do with roosters. These two unusual animals, both of whom belonged to a disconcerted clergyman, were observed to "mate" frequently, but the naturalists of the day doubted whether the union of the two was as complete as that of a rooster with a hen or a rabbit with another rabbit. So when the hen obligingly laid six normal-looking eggs, there was great excitement. What would hatch out?

Some people expected long-eared furry chickens to result; others, rabbits with beaks and feathers. But to the great disappointment of most, neither rabbit nor chicken nor anything in between emerged. The well-watched eggs merely sat and decomposed.

The eggs failed to develop because rabbits and chickens are different species of animals. As you may know, *species* may be defined as populations of organisms that retain their individuality in nature because they are reproductively isolated from other species around them. In general, reproductive isolation among animal species has two important and interrelated aspects: behavior and genetics.

The Parisian observers of the ill-fated "mating" of the rabbit and the hen were as familiar with the behavioral aspects of reproductive isolation as we are. They were well aware that animals of different species generally show no interest whatsoever in mating with one another. But what the Parisians did not realize was that, even if the behavioral aspect of reproductive isolation occasionally goes awry, the individuality of a species is still protected because species are genetically distinct from one another. What exactly does this mean?

Animals produce sex cells of two different types—eggs from the female and sperm from the male. The uniqueness of sexually-reproducing species (and this includes virtually all animals) may be thought of as having its basis in different blueprints, or genetic programs, for the elaboration of different species from fertilized eggs. As we shall see, each egg, as well as each sperm, normally contains half of the information necessary to set into motion the complicated process of elaborating a particular kind of animal according to the genetic program of the species to which the parents belong. But in order for proper development to occur, both sex cells that merge to form the fertilized egg must contain the same basic program.

When the Parisian rabbit and hen mated, it is not likely that their union resulted in a fertilized egg. This is because reproductive isolation operates even in sex cells, and rabbit sperm would be unlikely to penetrate and fertilize

The members of the Augustinian monastery in old Brno, Moravia (Czechoslovakia) in the early 1860s. Gregor Mendel, the discoverer of the basic patterns of heredity, is third from the right. (Photograph courtesy of Dr. V. Orel of the Moravian Museum, Brno.)

chicken eggs. In fact, even if we forced a rabbit sperm to fertilize a chicken egg by accurate mechanical injection of the sperm, the mating still would not produce feathered rabbits or long-eared chickens. A chicken egg fertilized in this way has received two conflicting programs, one for constructing a chicken and one for constructing a rabbit. And the programs for each are different enough that the artificially fertilized egg burns up its supply of intracellular fuel and then dies in the confusion of attempting to initiate the development of a composite creature from conflicting plans.

Actually, rare instances of successful interspecies mating do occur, not only among animals in experimental circumstances, but in nature too. (Also, interspecies crosses are much more common among plants than among animals.) Generally, such crosses occur only between species that are closely related by evolutionary descent, and that therefore presumably have similar genetic programs. Perhaps the most familiar animal issuing from an interspecies cross is the mule, the offspring of the mating of a female horse with a male donkey (Figure 1–1). Viable interspecies crosses also occur between closely related species of fish, of birds, and of porpoises, among others; however, offspring from such crosses are, like mules, usually sterile.

Differences in genetic programs between species are in large part responsible for the fact that animal species are usually morphologically distinct: that is, one can usually tell different species apart merely by looking at them. But then it is possible in many instances to distinguish animals from one another *within* a species, too (intraspecific variation). This is especially true of land-dwelling vertebrates, and is nowhere more obvious than in the human species, which is the most variable species known. Human beings have various skin colors ranging from almost pure white to jet black, have head and body hair ranging from perfectly straight to tightly kinked, and have unique fingerprints and faces—except for identical twins, who, as we shall see, have identical genetic programs. Yet within the human species, as in all others, the

1–1   The mule is a familiar hybrid produced by the mating of two closely related species—a (male) donkey and a (female) horse. Male mules are invariably sterile, but on rare occasions a female may become pregnant by a male horse or donkey. When this happens, the fetus is usually miscarried and born dead. Very rarely, the foal is carried for the full 10.5 months and is born alive. This occurs so infrequently that the Romans of classic times used the Latin phrase (*cum mule peperit*) meaning "when a mule foals" as the equivalent of our "once in a blue moon." (After "The Mule," by Theodore H. Savory, *Scientific American*.)

characteristics of individuals are not randomly distributed throughout the entire population. There are clear-cut geographic and racial differences as well as differences between related family lines.

The science of genetics is concerned with the study of heritable differences, both how they originate and how they relate to an individual organism's genetic program. Genetics also concerns itself with the biological basis of the transmission of traits in lineages and with the distribution of heritable characteristics within the populations that make up a given species. In chapters to come we will investigate the biochemical basis of heritable traits in individual persons and see how at least some of these traits may have come to have the distributions we observe within the human population today. But in this chapter our main concern is with identifying and explaining the simple patterns of inheritance shown by some rather clear-cut human characteristics that obviously run in families. For the most part, human patterns of inheritance can best be described by some basically simple—yet decidedly unobvious—principles first worked out in the 1860s by an Augustinian monk named Gregor Mendel (frontispiece).

What Mendel Did

Mendel discovered the basic patterns of inheritance by performing carefully planned experiments on the common garden pea, and his success was partly due to his wise choice of experimental subject. Pea plants are good subjects for simple genetic experiments for several reasons. First of all, individual pea plants have clear-cut differences in some easily recognizable alternate characteristics. For example, the ripe seeds may be either smooth or wrinkled, and either yellow or an intense green. Mendel chose to experiment with seven clearly alternate traits in his search for patterns in the way such traits are passed on from parents to offspring (Figure 1-2).

Another reason why pea plants make good experimental subjects is that they are self-fertilizing; that is, pea blossoms are so constructed that the male sex cells, which are contained in pollen grains, and the female sex cells, or eggs, are located in the same blossoms. (In other species, male and female sex cells may be produced separately, either by separate male and female flowers on the same plant or by flowers on separate male and female plants.) Self-fertilizing plants tend to breed "true," which means that their offspring usually resemble the parents exactly, at least in the alternate traits Mendel observed and recorded. True-breeding plants are thus good subjects for crosses of individual plants that differ in one or more alternate traits.

Mendel crossed plants that bred true for alternate traits and carefully recorded the distribution of these traits in their offspring. What he found is best illustrated by a recounting of some of his experiments.

Mendel began with two varieties of true-breeding pea plants, whose self-fertilized (self-pollinated) offspring had ripe seeds that were either round or wrinkled. He pinched off the pollen-producing parts (anthers) of each blossom on plants that produced only wrinkled seeds, and then fertilized the blossoms with pollen from plants tht bred true for round seeds. (He also fertilized some "round blossoms" with "wrinkled pollen" and produced es-

1–2   Mendel's genetic crosses of
the garden pea plant were mainly
concerned with the seven pairs of alternate characteristics shown here. A
given plant has one or the other of
each characteristic—not both.

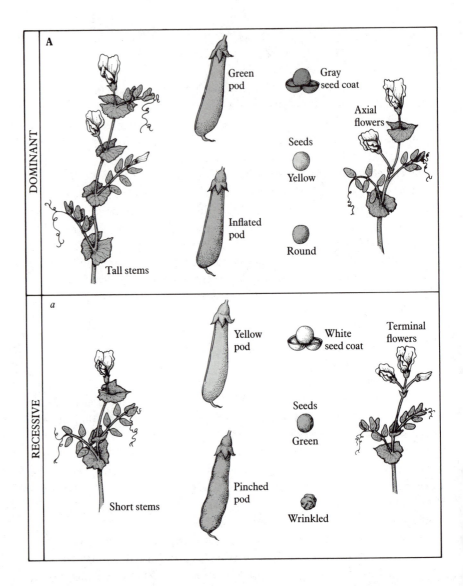

sentially the same results discussed in the following sentences.) Mendel then
tied little paper bags over the blossoms to prevent any wind-borne or insect-
borne pollen from contacting the artificially fertilized plants. When he opened
the pods of his experimental plants, he found that all of the seeds were round.
The alternate trait, "wrinkled," seemed to have disappeared in the first generation of progeny produced from the cross, the F_1 *generation*. Mendel then
planted the round seeds produced in the cross and allowed the resulting plants
to fertilize themselves, as they usually do. When he examined the seeds produced by the second generation (the F_2 *generation*), he sometimes found
round and wrinkled seeds lying together in the same pod (Figure 1-3). To
be more exact, he found that about 25 percent of the total number of seeds
in the F_2 generation were wrinkled. The trait "wrinkled," which had disappeared in the F_1 generation, had once again turned up in the F_2 generation
about 25 percent of the time.

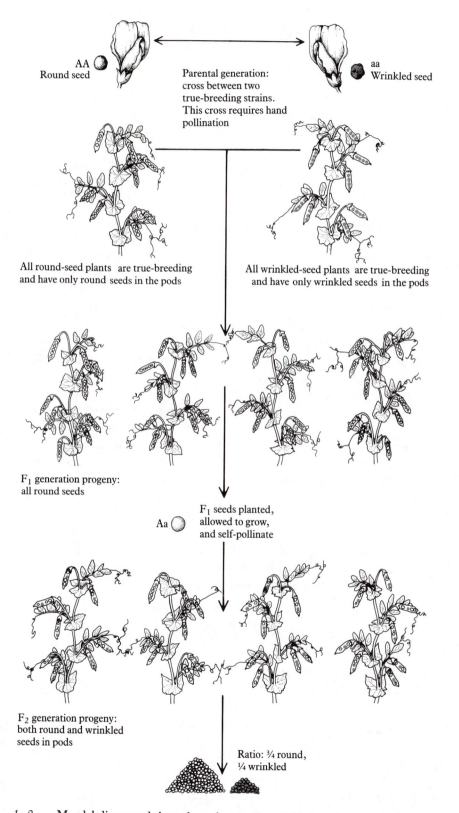

AA
Round seed

Parental generation:
cross between two
true-breeding strains.
This cross requires hand
pollination

aa
Wrinkled seed

All round-seed plants are true-breeding
and have only round seeds in the pods

All wrinkled-seed plants are true-breeding
and have only wrinkled seeds in the pods

F_1 generation progeny:
all round seeds

Aa

F_1 seeds planted,
allowed to grow,
and self-pollinate

F_2 generation progeny:
both round and wrinkled
seeds in pods

Ratio: ¾ round,
¼ wrinkled

1–3   Mendel discovered that when plants with wrinkled seeds are crossed with
plants with round seeds, all of the plants in the F_1 generation have round seeds.
Nonetheless, wrinkled seeds are regularly observed in members of the F_2 generation.

As shown in Table 1-1, Mendel found the same pattern for all seven of the traits he studied. For example, when true-breeding plants that had yellow seeds were crossed with those whose seeds were green, only yellow-seeded offspring were produced. Accordingly, Mendel called the member of the alternate pair of characteristics that showed up in all of the offspring of the F_1 generation, and in about 75 percent of the offspring in the F_2 generation, a *dominant trait*. And he named the trait that disappeared in the F_1, only to reappear in about 25 percent of the F_2, a *recessive trait*.

In order to explain the patterns he observed, Mendel proposed that inherited traits are transmitted from parents to offspring by means of independently inherited "factors" that are now known as *genes*. Furthermore, he found that he could predict the results of his experiments if he assumed that true-breeding lines of plants contributed either a dominant or a recessive factor to their offspring in the F_1 generation, and that members of the F_1 were therefore *hybrids*. That is, Mendel postulated that each member of the F_1 contained both dominant and recessive factors. This enterprising monk then invented a shorthand notation by which he could follow his hypothetical dominant and recessive factors through various lineages.

Mendel labeled the factor responsible for the dominant trait (round seeds) A, and he designated the factor responsible for the recessive trait (wrinkled seeds) a. (Geneticists still use uppercase letters to represent the genes responsible for dominant traits and lowercase letters to represent those responsible for recessive ones. Which letter is chosen to represent a given pair of alternate traits is arbitrary.) When both parents contribute an A to their offspring, the offspring are AA, and they produce only round seeds. In aa plants, which received an a from each parent, only wrinkled seeds are produced. Thus, true-breeding lines of plants are either AA (round) or aa (wrinkled). What happens if two such lines are crossed, as they were by Mendel to produce the F_1 generation? Clearly, all the offspring receive an A from one parent and an a from the other, so that all members of the F_1 must be Aa with respect to the alternate traits "round" or "wrinkled." What do Aa plants look like? Because A is dominant to a, all individuals in the F_1 will have round seeds. The trait "wrinkled" will seemingly have disappeared from

TABLE 1–1   MENDEL'S RESULTS FROM CROSSES INVOLVING SOME ALTERNATE CHARACTERISTICS OF THE COMMON GARDEN PEA.

Parent characteristics	F_1	F_2	F_2 Ratio
1. Round × wrinkled seeds	All round	5,474 round : 1,850 wrinkled	2.96 : 1
2. Yellow × green seeds	All yellow	6,022 yellow : 2,001 green	3.01 : 1
3. Gray × white seedcoats	All gray	705 gray : 224 white	3.15 : 1
4. Inflated × pinched pods	All inflated	882 inflated : 299 pinched	2.95 : 1
5. Green × yellow pods	All green	428 green : 152 yellow	2.82 : 1
6. Axial × terminal flowers	All axial	651 axial : 207 terminal	3.14 : 1
7. Long × short stems	All long	787 long : 277 short	2.84 : 1

the F_1, just as Mendel observed. However, the factor responsible for the recessive trait has not, in fact, disappeared; its effects are simply masked by the presence of factor A and, in later crosses, the effect of a can become obvious once again.

Consider what happens when the hybrid plants of the F_1 are allowed to self-fertilize and to produce offspring. All of the parents are Aa, and can contribute either A or a to their offspring, and in fact do so in equal proportions. About half the offspring get A and half get a from *each* parent. This means that three different kinds of offspring can result: AA or aa plants if both parents happen to contribute the same factor, and Aa plants if each parent happens to contribute a different factor.

An easy way to predict what will happen in a given cross is to construct a table that allows us to keep track of all possible combinations of factors. Across the top of the table are listed the factors that one parent can contribute; those from the other parent are listed down the left-hand side. In our example, both parents are hybrids, and so both can contribute either A or a. This can be represented diagrammatically as follows:

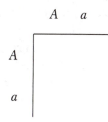

Then by simply drawing in the boxes and combining the factors, we generate the following table. (Uppercase letters are always written first.)

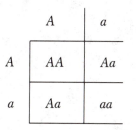

	A	a
A	AA	Aa
a	Aa	aa

Thus, the combinations we should expect in the offspring are AA, Aa, and aa. Notice that Aa appears in the table twice. This means that about two out of every four offspring will be Aa. Or, more generally, about 50 percent of the offspring will be Aa. Similarly, about 25 percent of the offspring will be AA and the remaining 25 percent aa. Looking at it another way, 75 percent of the offspring are either Aa or AA and therefore have round seeds, and 25 percent are aa and have wrinkled seeds. These ratios are exactly those observed by Mendel in the F_2 generation.

Exactly? Not quite, because whenever one is dealing with large numbers of objects that behave independently and are expected to exist in a predicted ratio, what is actually observed is not the exact theoretical ratio but something close to it. Assuming that the theory is correct, how close the observed ratio is to the theoretical one is a matter of chance. The observed ratio is usually somewhat higher or lower than the predicted one; only rarely will chance dictate that the observed and predicted ratios be the same or very close. In 1936, the American statistician and geneticist R. A. Fisher (see Figure 7-1) carefully analyzed Gregor Mendel's published ratios of dominant to recessive traits in the F_2 generation and made a surprising discovery: they are simply too good to be true. That is, Mendel's published ratios show much less variation on either side of the anticipated 3:1 ratio than the laws of probability would predict. Does this mean that Mendel fudged his data, or at least polished them a little? There is no reason to suspect the monk of out-and-out cheating, but it seems likely that some kind of conscious or unconscious bias affected his observations. Figure 1-4 provides a clue to what the nature of the bias may have been. As you can see, the difference between "round" and "wrinkled" peas is usually clear-cut, but there are some borderline cases. Mendel may have unconsciously placed such relatively rare, borderline specimens in one or the other category according to what his theory would predict. In support of this idea is the fact that Mendel, unlike other persons working with the same species of pea plant, never once came across a pea he couldn't unequivocally classify as either "round" or "wrinkled."

Using the same kind of reasoning that had correctly predicted the results of crosses of a single pair of alternate traits, Mendel predicted what would happen if he crossed plants differing in *two* alternate characters. He crossed plants that had round yellow peas (both dominant) with plants bearing wrin-

1–4   Round peas and wrinkled peas (the "seeds" recorded in Mendel's crosses) are usually easy to tell apart, but some specimens are difficult to classify.

kled green peas (both recessive) and, as predicted, all members of the F_1 were round and yellow. He then allowed the F_1 hybrids to self-fertilize. If the factors he postulated did indeed exist and behave as independent units, then he expected to find four kinds of peas in the F_2: round yellow, wrinkled yellow, round green, and wrinkled green. Moreover, he predicted that he would find them in the ratio $9:3:3:1$. He performed the crosses and found the actual ratios to be as he predicted, allowing for small deviations introduced by chance (Figure 1-5).

Mendel had discovered the most fundamental patterns of inheritance, and they have stood the test of time to the present day. We shall soon see how they apply to human lineages. He published his results in 1866, but for the most part his manuscript was ignored. By and large this was because Mendel's factors could not be seen; they were rather mysterious units for which no physical basis was known. But all that had changed when Mendel's work was rediscovered in 1900, when it was fully appreciated for the first time. What made the difference was that in the interim biologists had discovered what they presumed to be the physical basis of Mendel's mysterious factors. They had discovered chromosomes.

Chromosomes and Mendel's Patterns

When Mendel published his results in 1866, it was well known that cells are the basic building blocks of all living things. But at that time these fundamental units were poorly known and largely undescribed, because the manufacturing of microscopes and the preparation of specimens for microscopic study had not yet become highly developed arts. Nonetheless, it was known that most plant and animal cells have a distinct nucleus inside them, and that within the nucleus are rodlike structures called chromosomes. In general, chromosomes are clearly visible only in cells that are in the process of dividing. As we discuss later in Chapter 3, in resting, nondividing cells the chromosomes are still inside the nucleus, but they are much thinner and are highly entangled with one another so that individual chromosomes cannot be distinguished.

By the time Mendel's work was rediscovered in 1900, enough was known about the remarkable behavior of chromosomes in dividing cells to suggest that the observed patterns of inheritance could be explained in cellular terms by assuming that Mendel's independent factors were located on chromosomes. The gist of the evidence, as first described about the turn of the century, is this: within the nucleus chromosomes exist in pairs, except in a single, revealing instance. The exception is the sex cells, whose nuclei contain only one member of each pair, or half the number of chromosomes in other body cells. How does this tie in with the inheritance of alternate traits described by Mendel?

Assume that the dominant and recessive forms of a particular factor are located one each on the two chromosomes of a particular pair. Thus, an Aa individual has an A on one chromosome of a given pair and an a on the other. When the individual produces sex cells, pairs of chromosomes separate so that each egg or sperm contains either an A chromosome or an a chromosome.

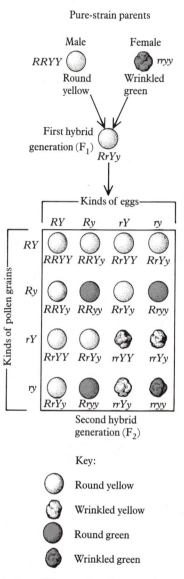

1–5 The results of a cross between pea plants that differ in two pairs of alternate characteristics. The traits "round" and "yellow" are dominant, whereas "wrinkled" and "green" are recessive. On the average, out of every 16 seeds produced 9 are round and yellow, 3 are wrinkled and yellow, 3 are round and green, and only 1 is wrinkled and green.

Then, when self-fertilization occurs, pairs of chromosomes are reunited once again, and the resulting offspring are either *AA, Aa*, or *aa*, depending on which factors happened to be located on the particular chromosome pairs that were reunited.

You will recall that Mendel followed the patterns of inheritance of seven pairs of alternate traits of the common garden pea. It turns out that pea plants have seven pairs of chromosomes, which correlates with Mendel's observation that all seven of the traits he studied behaved independently. That is, Mendel's factors showed *independent assortment* because the traits that he studied are determined by factors located on different pairs of chromosomes. (Later reserach revealed that at least two of the traits that Mendel studied are actually determined by genes located on the same chromosome. But the genes are located at nearly opposite ends of the chromosome, and, for reasons discussed later, they therefore tend to sort out in the sex cells as if they were located on different chromosomes.)

It is now known that the number of chromosome pairs normally present in the nucleus can vary widely from species to species. (Remember that only one member of each pair is present in an animal's sex cells.) It is also known that each chromosome pair usually carries factors responsible for many different traits.

How does all of this information relate to people? Our discussion of Mendel's work provides a background for discussing human chromosomes and for relating them to the patterns of inheritance shown by some alternate traits in human families. But before we go any further, we should introduce some terms that describe the genetic makeup of an individual, human or otherwise. Familiarize yourself with these words now, for they will be used repeatedly in the discussion that follows.

Some Definitions

Mendel's inherited factors, the units of heredity, are now called *genes*, and the two or more forms of the genes responsible for alternate traits (*A* and *a* in our example) are called *alleles*. We will discuss the biochemical basis of genes in Chapter 3. For now it is enough to know that many genes are associated with so-called gene products, which are usually complex molecules that participate in the body's intricate web of biochemical reactions.

Individuals in whom the two alleles of a given pair are the same (*AA* or *aa*) are called *homozygotes*, whereas *heterozygotes* are individuals in whom the two alleles of a given pair are different (*Aa*).

Recall that you cannot distinguish *Aa* heterozygotes from *AA* homozygotes merely by looking at them. (Both have round seeds.) But the two can be told apart if they are crossed with known heterozygotes (*Aa*) or with known *aa* individuals. Thus, if wrinkled seeds (*aa*) turn up in the offspring of a cross of two round-seeded pea plants, we can conclude that both parents must have been *Aa*. If only round seeds are produced, then the offspring are either *Aa* or *AA*, and the parent crossed with the known heterozygote must have been *AA*. To distinguish individuals that look alike but nonetheless have different genetic constitutions, geneticists use the terms *phenotype* and *gen-*

otype. Phenotype is a description of what an individual looks like, and genotype describes the individual's genetic constitution. In our example, individuals of phenotype "round" may be either of two genotypes, *Aa* or *AA*.

With these terms in mind, let us discuss the chromosomes of the human species and relate them to some fairly obvious patterns of inheritance in human families.

Human Chromosomes—A First Look

Although the existence of chromosomes has been known for over a hundred years, the exact number of chromosomes that characterizes the human species was not discovered until 1956. This seems extraordinary, but in fact, the chromosomes of most mammals are not only numerous but difficult to prepare for detailed microscopic study; only relatively recently have satisfactory and consistent methods for visualizing them been developed. Human cells contain a total of 46 chromosomes in 23 pairs. That is, body cells, or *somatic cells*, contain 46 chromosomes in their nuclei, and sex cells contain only a single member of each pair, or a total of 23.

Human chromosomes can be made visible for detailed study by means of the techniques outlined in Figure 1-6. First, a blood sample is collected (a few drops are enough) and placed in a test tube containing a liquid that has the same concentration of salt as human blood. The red blood cells are allowed to settle to the bottom of the tube and are then removed. A chemical is then

1–6   The techniques used in preparing human chromosomes for microscopic study.

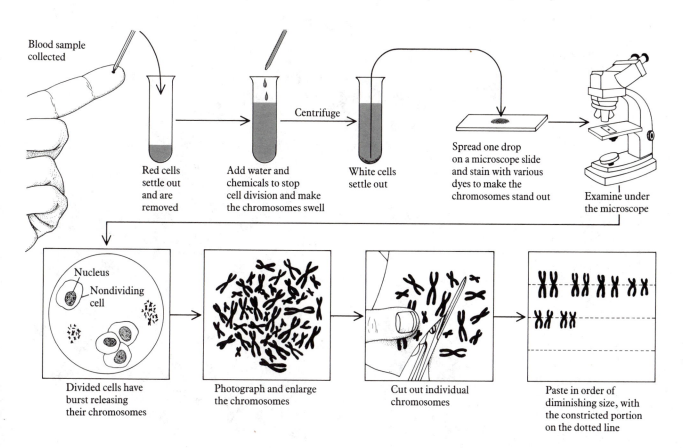

Blood sample collected

Red cells settle out and are removed

Add water and chemicals to stop cell division and make the chromosomes swell

Centrifuge

White cells settle out

Spread one drop on a microscope slide and stain with various dyes to make the chromosomes stand out

Examine under the microscope

Nucleus

Nondividing cell

Divided cells have burst releasing their chromosomes

Photograph and enlarge the chromosomes

Cut out individual chromosomes

Paste in order of diminishing size, with the constricted portion on the dotted line added to stimulate the white blood cells to divide, thereby making their chromosomes visible. After two or three days, when a sufficient number of white cells has accumulated, water and more chemicals are added to make the chromosomes swell and to stop cell division at a stage when they are most easily distinguished from one another. The dividing cells are then broken open, spun in a centrifuge in order to concentrate them, placed on a microscope slide, and stained with dyes that are selectively taken up by the chromosomes. When the specimen is examined under the microscope, intact, nondividing cells and groups of chromosomes that were released from the disrupted, dividing cells are seen. A close-up of the complete set of chromosomes of a normal man is shown in Figure 1-7. The best group of chromosomes is carefully searched for, photographed, and enlarged. The individual chromosomes are then cut out of the enlarged photograph like paper dolls and lined up in matching pairs of diminishing size to form a *karyotype*. The karyotype of a normal man is shown in Figure 1-8. Note that the chromosomes vary not only in size but also in shape. Some are X-shaped, with arms of nearly equal length; others have arms of very unequal lengths. Also note that all chromosomes have a *primary constriction*, or *centromere*, from which the arms extend.

As shown in Figure 1-8, the 23 pairs of human chromosomes are placed in one of seven groups designated by the letters A through G. Of the 23

1–7   The chromosomes from a disrupted, dividing white blood cell of a normal man, as seen under the light microscope.

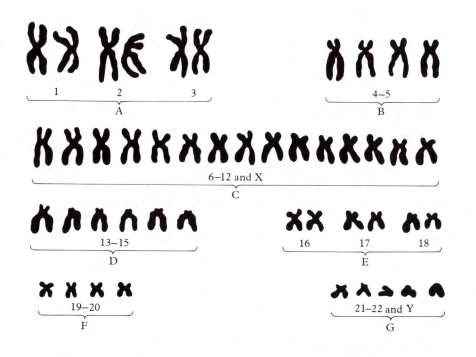

1–8 The karyotype of a normal human male. The chromosomes have been photographed, enlarged, cut out, and arranged in groups according to size and shape.

pairs, 22 are for all practical purposes perfectly matched in both sexes and are called *autosomes*. The remaining pair are called *sex chromosomes*, and though the members of this pair are apparently identical in women, they are not identical in men. With regard to genotype, women are said to be of sex chromosome constitution XX, whereas men are said to be XY. Sex chromosomes will be discussed in the following chapter, and the fine structure of human chromosomes will be covered in Chapter 5. For now, we turn our attention to the inheritance of some human traits determined by genes located on autosomes. But first, it is helpful to elaborate a little on exactly how the terms *dominant* and *recessive* apply to human characteristics.

Most of what is known of human genetics has been learned by studying various kinds of diseases or abnormalities that obviously run in families. In the final analysis, all gene-dependent differences among human beings are differences between physiological processes that occur inside cells. Sometimes the way in which genetic differences between cells can result in phenotypic differences between individuals is obvious. For example, albinos are lightly pigmented because their cells are unable to synthesize the dark-colored pigment melanin properly. But often it is not at all obvious how phenotypic abnormalities relate to gene-dependent abnormalities in cellular physiology and in biochemistry. For example, it is not obvious how the gene for six-fingered dwarfism (see the following discussion) brings about its effect. Moreover, the exact biochemical or physiological defect that is associated with a genetically determined abnormality is known for only about one trait in five.

Many heritable human disorders, most of them individually rare, are the result of a single abnormal allele for which an affected person may be either homozygous or heterozygous. These rare disorders therefore have simple Mendelian patterns of inheritance, as we are about to discuss. An abnormal allele is dominant or recessive depending on whether its effects are evident in a single dose (in heterozygotes) or whether the allele must be present in a double dose (in homozygotes) to produce its effects.

On the other hand, many common disorders, such as high blood pressure and some relatively frequent congenital malformations such as cleft lip and palate, do not have simple Mendelian patterns of inheritance. This is because these abnormalities, like many others, result from the interaction of many genes and many nongenetic environmental factors. As we will discuss in Chapter 8, this is also true of many normal human characteristics such as height and intelligence. *In the discussion of the patterns of inheritance shown by autosomal dominant and recessive traits that follows, it should be borne in mind that the existence of rare abnormal alleles in affected persons implies the presence of normal alleles in normal persons.* The study of genetic abnormalities can thus help us to get an idea of how extensive the normal human genetic program really is.

Autosomal Dominant Inheritance

At least 1000 human traits are known to have their genetic basis in dominant genes located on autosomes and about 1000 additional traits are suspected, but not proved, to have this pattern of heredity. As you know, Mendel discovered that dominant traits are manifested both by heterozygotes (*Aa*) and by homozygotes (*AA*). But almost all human beings who manifest documented autosomal dominant traits turn out to be heterozygotes. This is because dominant genes for the most part produce undesirable effects. That is, persons manifesting dominant traits are usually at some kind of disadvantage

1–9   A Norwegian family, some of whose members have woolly hair. (From Mohr, *Journal of Heredity*, *23*, 1932.)

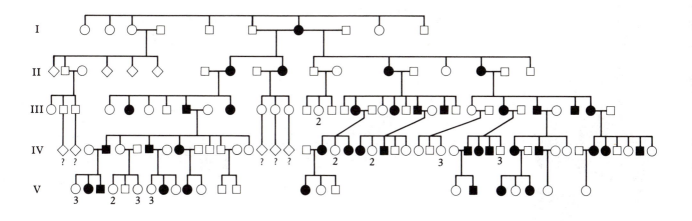

compared to their normal peers. Apparently, homozygotes for dominant traits are at such a disadvantage that most of them do not survive life before birth and die as embryos. Besides, autosomal dominant traits are rare to begin with, so it is unlikely that two affected persons would come together to produce homozygous offspring. Thus, we can usually assume that persons manifesting autosomal dominant traits are heterozygous.

A good example of autosomal dominant inheritance is provided by the benign trait known as "wooly hair," whose distribution has been well documented in Norwegian families. Affected persons have hair that is tightly kinked and very brittle, so that it breaks off before growing very long (Figure 1-9). As usual, persons manifesting this dominant trait are heterozygous, and their genotype can be symbolized as Ww (W for wooly). If an affected person (Ww) and an unaffected, normal person (ww) produce offspring, we would expect about half of them to have normal hair and half to have wooly hair. Moreover, if the gene that determines the trait is located on an autosome, then the affected offspring should include roughly equal numbers of males and females, and the trait should be transmitted from either parent to both sons and daughters. That this is true is shown in Figure 1-10, which is human pedigree outlining the transmission of wooly hair through several generations. The symbols used in outlining pedigrees are these: women are presented by circles and men by squares. Fetuses stillborn before their sex could be determined are indicated by diamonds, and deceased individuals are indicated by a diagonal slash through the symbol. Individuals directly connected by horizontal lines mate, and their offspring are indicated at the end of short vertical lines. Affected individuals are indicated by blacking in the symbols. For example, suppose that a normal man marries a wooly-haired woman and that they produce four offspring, two boys and two girls, one affected and one normal child of each sex. This is symbolized as follows:

1–10 A pedigree showing the transmission of woolly hair through five generations. The symbols are explained in the text. If more than one offspring are represented by a symbol, the number represented is given below the symbol.

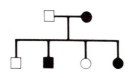

1–11　　Abraham Lincoln's unusual body proportions, his "wandering eye," and other supposed anatomical peculiarities have led some physicians to believe that he was mildly afflicted by Marfan's syndrome. (Photograph courtesy of The Louis A. Warren Lincoln Library and Museum.)

If the affected daughter then marries a normal man and produces five offspring, including three normal daughters and two affected sons, the diagram is extended to:

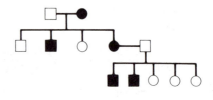

Pedigrees of families including wooly-haired individuals have also been recorded in Holland and the United States. If we pool all the data concerning the offspring from marriages between one affected and one normal person, we find, as we would expect, that within the limits of chance half of the sons and half of the daughters inherit the trait.

A fairly constant feature of most autosomal dominant traits in human pedigrees is that they can vary widely in severity from one individual to another. The degree of severity is referred to as the *expressivity* of the trait. A good example of variable expressivity in human pedigrees is the condition known as Marfan's syndrome. (A *syndrome* is a group of signs and symptoms that are present at the same time and characterize a particular abnormality. Syndromes are frequently named after the person who first described them.) Marfan's syndrome is an autosomal dominant disorder that is manifested by elongated hands and feet, abnormalities in the placement of the lens of the eye, a long, narrow skull, an overgrowth of the ribs, and many other abnormalities. An individual with Marfan's syndrome may manifest one, a few, or all of these abnormalities. Because of his distinctive appearance and family background it has been suggested that Abraham Lincoln may have been mildly afflicted by Marfan's syndrome. The biography written by his lifelong friend and law partner, William H. Herndon mentions that the famous president's head was "long and tall" and that "his legs and arms were abnormally, unnaturally long and in undue proportion to the remainder of his body." Moreover, Herndon observed that "it was only when he stood up that he loomed over other men" (Figure 1-11). Lincoln's peculiar body proportions and the diagnosis of Marfan's syndrome in a descendant of his great-great-grandfather, Mordacai Lincoln II, have been offered as evidence that the president had Marfan's syndrome, though with a low degree of expressivity. Against the diagnosis are Lincoln's general robustness, at least in his youth (in the same biography Herndon says: "He could strike with a maul a heavier blow—could sink an axe deeper into wood than any man I ever saw"), and the absence of any single characteristic or combination of characteristics considered to be diagnostic of the disease.

By and large, whether or not a particular autosomal dominant gene is fully expressed depends on the rest of the person's genes. In other words, the presence of certain other genes or combinations of genes can markedly influence whether or not any trait is manifested, and this is especially true of autosomal dominant ones. The proportion of individuals of a given genotype who actually manifest the trait associated with the genotype is known as the

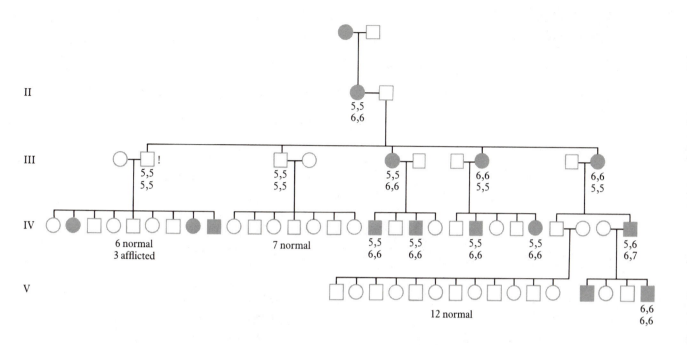

penetrance of the trait. Penetrance differs from expressivity in that it is an all-or-none phenomenon. A good example of penetrance in human pedigrees is *polydactyly*, the presence of more than the usual number of fingers or toes. As shown in Figure 1-12, certain persons who carry the dominant gene for polydactyly do not express it but can nonetheless pass the gene on to their offspring, who do have extra fingers or toes.

Another feature of autosomal dominant traits is that they sometimes appear unexpectedly among the offspring of unaffected parents. In Marfan's syndrome, this occurs about 15 percent of the time. Those who are affected thereafter pass the trait on to about half of their offspring, as expected. In general, the sudden, unexpected appearance of an autosomal dominant trait in a lineage from which it was not previously known is the result of a *mutation*. In our example, a heritable change occurs spontaneously within the genetic material and thereby transforms a normal gene into the one responsible for Marfan's syndrome. (Mutations will be discussed further in the following chapters.)

Codominance: The ABO Blood Group

So far we have been discussing autosomal dominant traits that depend on a single pair of alleles, which is to say that the genes determining such traits come in only two forms, *A* and *a*. As we have seen, most persons who manifest autosomal dominant traits are heterozygous for the allele in question. However, *multiple alleles* are also known to influence the inheritance of some well-known and important human characteristics, including the determination of a person's *blood group*. There are many blood groups, most of which depend on different sets of alleles. Perhaps the best-known set of multiple alleles is that determining the ABO blood group, which is important enough to be worth discussing further.

1–12   The two girls shown here have inherited the dominant gene that results in extra fingers or toes (rarely, both). Note that the man on the left in the third generation of the pedigree (above) had a mother with six toes, and that he and his normal wife produced six normal and three polydactylous children. In this man, the gene for polydactyly was not manifested. (Photograph courtesy of Wide World Photos.)

Three alleles determine to which ABO blood group a person belongs. These can be symbolized as I^A, I^B, and I^O. These alleles are always found at a particular place, or *locus*, on a particular pair of autosomes, and any one person has two out of the three alleles. $I^A I^A$ individuals are of blood group A, as are $I^A I^O$ individuals. Similarly, persons who are $I^B I^B$ or $I^B I^O$ are group B. To complete the possibilities, $I^A I^B$ individuals are of blood group AB and $I^O I^O$s are of group O.

Both of the alleles I^A and I^B are dominant to I^O. Moreover, when I^A and I^B occur together, both have an effect on the surface of red blood cells (and both therefore stimulate the production of antibodies). Thus, I^A and I^B are said to show *codominance*.

ABO blood grouping has been carried out nearly world wide for several reasons. First, ABO compatibility is essential in performing blood transfusions. Second, ABO groups vary enough from one group of people to another that their patterns of variation have been useful to physical anthropologists in the study of human races (see Chapter 7).

An unintended but useful side effect of widespread ABO typing has been its application to cases of disputed paternity. Consider the following example. During the course of divorce proceedings a woman of type A (genotype $I^A I^A$ or $I^A I^O$) seeks child support from a man of type O (genotype $I^O I^O$), claiming that he is the father of her recently born child. If it should turn out that the child is of type AB (genotype $I^A I^B$), the case would be thrown out of court, because the supposed father is capable of contributing only the allele I^O to his offspring. The mother can contribute only I^A or I^O, so the allele I^B must have been contributed by someone else. (The likelihood of a mutation having occurred in one of the parents' sex cells is negligible.) However, if the presumed father were of type B (genotype $I^B I^B$ or $I^B I^O$), then overall there would be a 50 percent chance that he could in fact be the father and the trial would continue.

A famous case of disputed paternity that should have been settled because of the blood types of the mother, child, and purported father involved the comedian Charlie Chaplin. In 1944 Chaplin, who was of blood type O, was sued for child support by the actress Joan Barry, who was of blood type A. Because the baby was found to be of blood type B, Chaplin could not have been the father. Nonetheless, the court ordered the comedian to pay child support. Why? Because at that time data concerning blood types were not admissible evidence in California, and they provided the only evidence that Chaplin was not the father.

We now turn our attention to the inheritance of some autosomal traits whose expression implies that an affected person, unlike most people who are afflicted by a dominant trait, is a homozygote.

Autosomal Recessive Inheritance

At least 600 human traits are definitely known to have an autosomal recessive pattern of inheritance, and nearly 800 additional traits are suspected but not proved of being transmitted in this way. As first discovered by Mendel, heterozygous individuals (*Aa*) do not manifest autosomal recessive traits because of the masking effect of the dominant allele. Thus, persons who manifest autosomal recessive traits are generally *homozygous recessive*; that is, their genotype is *aa*.

Compared with their dominant counterparts, autosomal recessive traits are likewise usually associated with diseases or abnormalities, and they are not quite so rare. This is because most persons who are heterozygous for an autosomal recessive trait (of genotype *Aa*, also called *carriers of the trait*) are not at much of a disadvantage compared to normal (*AA*) individuals, so the trait may become widely disseminated even if affected homozygous persons choose not to reproduce. Also, those affected by autosomal recessive traits tend to be less variable than those affected by autosomal dominant ones. That is, the expressivity of autosomal recessive traits is about the same for all those who are affected.

A good example of autosomal recessive inheritance is the condition known as *albinism*. Albinism is one of the most common and widespread genetic disorders. Affected individuals include not only human beings of all races but also other mammals, insects, fish, reptiles, amphibians, and birds. Albinism results from the body's inability to synthesize the dark-colored pigment *melanin* properly. Melanin is the principal pigment that imparts color to human skin, hair, and eyes, so human albinos generally have white hair and pink or only lightly colored irises (Figure 1-13). Because of their lack of pigment, the skin and eyes of albinos are abnormally sensitive to the effects of sunlight, and because of their unusual appearance, human albinos may receive special treatment from other members of their species. For example, the Aztec emperor Montezuma is said to have included many albinos among the members of his "museum" of living human "curiosities," and albinos among the present-day San Blaz Indians of Panama (known as "moon children" because they avoid bright sunlight) are not permitted to marry.

Until quite recently it was thought that albinism was the result of a single, specific defect in the synthesis of melanin from the amino acid tyrosine. (In later chapters we will discuss the biochemistry of melanin synthesis as it relates to inborn errors of metabolism.) But then a well-documented pedigree of albinism in England showed that two albino parents produced four children, none of whom were albinos. As shown in Figure 1-14, the mating of parents manifesting the same autosomal recessive trait can produce only affected offspring, because each of the parents contributes an *a*. How then can the pattern observed in this unusual English family be explained?

It turns out that there are at least two, and perhaps as many as six, genes that result in albinism, depending on where the biochemical block in the synthesis of melanin occurs. And all of these defects in melanin synthesis are inherited as autosomal recessive traits. Therefore, it is possible for two albinos who are homozygous recessive for different defects in melanin synthesis (and who are therefore albinos for different reasons) to produce normal offspring, as shown in Figure 1-14.

In the United States, about one white person in 38,000 and one black person in 22,000 are albinos. But circumstances are known in which the percentage of albino offspring produced is higher than that of the population at large. The best known example is that of marriages between relatives, or

1–13 Two albino parents and their albino daughter. (After Davenport, *Journal of Heredity*.)

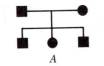

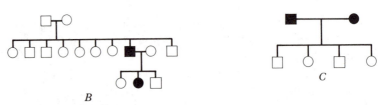

A

B

C

1-14   A, the mating of albino parents almost always results in all albino offspring. B, a pedigree showing albinism among the offspring of an albino father and an unaffected carrier mother. C, a pedigree of albinism in which albino parents produced four nonalbino offspring.

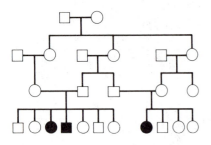

1-15   A pedigree showing marriages between cousins that resulted in albino offspring. The consanguineous matings occurred in the third generation.

consanguineous marriages. Although only about 0.1 percent of marriages in the United States are between first cousins, about 8 percent of albino children result from first-cousin marriages. (First cousins are the offspring of brothers and sisters who married unrelated spouses. See Figure 1-15.) How does the incidence of albinism, or any other autosomal recessive trait, relate to the degree to which an affected person's parents are related?

Most autosomal recessive traits are rare. Nonetheless, the brothers and sisters of someone who carries an allele for a rare autosomal recessive trait are very likely to be carriers too. (If a person carries an allele for albinism, the person's normal brothers and sisters have a 50 percent chance of also carrying the allele, because at least one of their parents must be a heterozygote. Work this out for yourself.) Similarly, persons descended from a common ancestor known to have manifested or carried a particular trait are also more likely to be carriers. Thus, persons manifesting rare autosomal recessive alleles tend to cluster in certain family lines because the mating of rather closely related individuals is likely to bring two rare autosomal recessive alleles together to produce an affected individual.

One example is the autosomal recessive condition known as *six-fingered dwarfism* (the Ellis van Creveld syndrome), which is unusually frequent among the Old Order Amish of Lancaster County, Pennsylvania (Figure 1-16). Amish people usually choose to marry other Amish people, and because their numbers are relatively small to begin with, this means that marriages between individuals who have common ancestors occur frequently. Accordingly, whereas six-fingered dwarfs are very rare elsewhere, they exist in at least 33 Amish families. Apparently, one of the original founders of the sect in eastern Pennsylvania was a heterozygous carrier of the gene for six-fingered dwarfism. The gene has since become widespread in the population and is frequently found in the homozygous state because of consanguineous marriages.

In general, the more closely related two persons are, the greater the chance that their offspring will manifest some (usually detrimental) autosomal recessive trait. This is best illustrated by data about the offspring of incestuous unions of fathers and daughters and brothers and sisters. Information on the offspring of 31 such unions is available from England and the United States. Six of the offspring died early in life and twelve were severely affected physically or severely retarded mentally. Only 42 percent of the children were apparently normal. (Almost all human societies have strict cultural prohibitions against incestuous unions. Is this because the deleterious genetic consequences of such unions are so obvious? Almost certainly not. Rather, it appears that incest taboos are outgrowths of social and cultural factors, and not the result of a primitive form of applied genetics.)

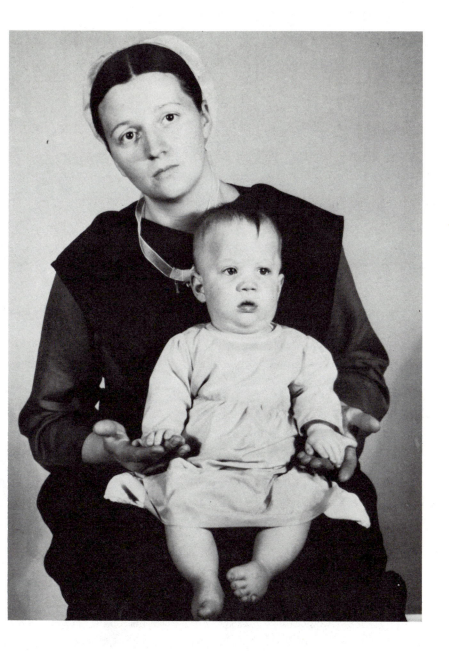

1–16 An Old Order Amish mother and her child, who has six-fingered dwarfism. (Photograph courtesy of Dr. Victor A. McKusick.)

Autosomal recessive traits are of particular interest in genetic counseling. Most often the relatives of an affected person seek advice about whether they are carriers, or about the chances of their having an affected child. It is sometimes possible to infer a person's genotype directly from pedigree studies, and the person can thereby be told his or her status as a carrier. But even when a person'a genotype cannot be deduced by pedigree analysis, it is often possible to determine it by performing various physiologic and biochemical tests. Some heterozygotes for autosomal recessive traits do indeed manifest the fact that they harbor both dominant and recessive alleles.

We have already mentioned that homozygous individuals who manifest autosomal recessive traits usually do so about equally. The effects of a dominant allele generally mask the effects of a recessive one in heterozygotes, but

1–17   A company of 175 soldiers arranged in groups according to height. The lower row of numbers indicates height in feet and inches, the upper row the number of men in each group. (From Blakeslee, *Journal of Heredity*, 5, 1914.)

| 1 | 0 | 0 | 1 | 5 | 7 | 7 | 22 | 25 | 26 | 27 | 17 | 11 | 17 | 4 | 4 | 1 |
| 4:10 | 4:11 | 5:0 | 5:1 | 5:2 | 5:3 | 5:4 | 5:5 | 5:6 | 5:7 | 5:8 | 5:9 | 5:10 | 5:11 | 6:0 | 6:1 | 6:2 |

in some instances it is possible to infer the presence of the recessive allele by subjecting the suspected carrier to some kind of stress. For example, heterozygous carriers for sickle-cell anemia (whose biochemistry and genetics will be discussed in Chapter 3) can be detected if their blood is subjected to lower-than-normal concentrations of oxygen. Under this stress, some of the carrier's red blood cells take on the characteristic sickle shape, and thus reveal heterozygosity.

In general, it is possible to devise some sort of measurement that will reveal heterozygous carriers of autosomal recessive traits, as long as the physiological or biochemical defect in question is known. But the exact defect is known for only about one genetically determined abnormality in five. Besides, the inheritance of many traits of normal human beings does not fit into any of Mendel's patterns, in spite of the fact that the traits obviously have some sort of genetic basis. This is especially true of traits that are not clearly alternate, but rather are continuously distributed throughout the population. Such traits do not manifest themselves as sharply defined pairs of phenotypes like "round" or "wrinkled," but rather as continuous gradations. For example, normal body height varies widely but in general is continuously distributed in a given population (Figure 1-17). Even though there is a definite tendency toward tallness or shortness in families, there may be widespread differences in the heights of parents and offspring. Also, as with all continuously varying traits, height is known to be affected, not only by a person's genes, but by the environmental conditions in which the person grows up and lives.

It was Mendel himself who first proposed an explanation for the inheritance of continuously graded traits. Based on some experiments he performed on bean plants with colored and white flowers, he suggested that *more than one pair of genes* transmitted such traits. For the most part, Mendel's explanation has stood the test of time, though we now are much more aware of the effects of the environment on the expression of continuously varying traits than Mendel was. We will return to the inheritance of continuously varying traits and to the effects of the environment on the expression of a person's genotype in later chapters. But for now, let us turn out attention to some human conditions that result from the presence of abnormal numbers of autosomes.

Down's Syndrome and Other Abnormalities in the Number of Autosomes

People who have Down's syndrome have surely existed since ancient times, but the condition was not described in detail until 1866. Those who are affected with it are short in stature, frequently have serious malformations of the heart, and usually have characteristically shaped heads with distinctive eyelids and faces, among many other features. More important, persons who have Down's syndrome are almost without exception severely mentally retarded. (To the Europeans who first described the condition, the characteristic appearance of the eyelids suggested the facial features of Mongoloid peoples, and the condition was referred to as "mongolism" or "mongoloid idiocy." In fact, the eyelids of persons who have Down's syndrome are quite different from those of persons belong to Mongoloid races, and the earlier, inaccurate terminology has therefore been dropped.)

Down's syndrome occurs sporadically. That is, affected individuals are usually the offspring of normal parents. After it was known that the incidence of the disease varies directly with the age of the mother at the time of birth (as discussed later, older mothers have a greater tendency to produce affected offspring), it was assumed that the syndrome was produced by an unfavorable interaction between mother and fetus during the course of pregnancy. Then, in 1959 (shortly after techniques for seeing human chromosomes had been perfected), French investigators discovered that the somatic cells of those who had *Down's syndrome* contained 47 chromosomes, one more than usual. As shown in Figure 1-18, the extra chromosome is a rather small autosome belonging to "group G," and by convention it is designated chromosome-21.

The presence of three copies of a given chromosome is known as *trisomy*, and individuals who are trisomic for a human chromosome tend to have many

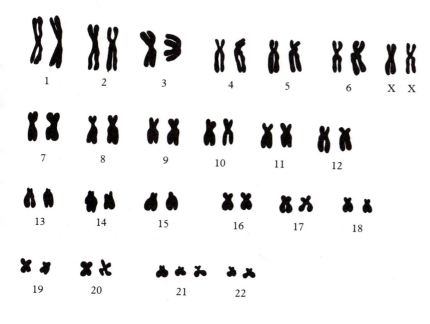

1–18 Down's syndrome most often results from the presence of an extra copy of chromosome-21, as in the karyotype shown here.

biochemical and physical abnormalities. The presence of an extra chromosome indicates the existence of extra copies of all of the genes located on the chromosome. This degree of genetic imbalance has major consequences during embryonic development, when the rate of cell division is high, and most trisomic fetuses undergo spontaneous abortion (miscarriage) in early pregnancy.

How do the somatic cells of persons who have Down's syndrome end up with an extra chromosome-21? In brief, trisomy-21 (and trisomy for other chromosomes) is usually the result of an accident that occurs during cell division. What happens is this: duplicated pairs of chromosomes fail to separate during cell division. Some sex cells receive two chromosomes-21 and some receive none. This failure of chromosomes to sort out properly during cell division is called *nondisjunction* (Figure 1-19).

What happens if nondisjunction occurs during the production of a human egg that is then fertilized by a normal sperm? With regard to Down's syndrome, there are two possibilities. If the abnormal egg contains two chromosomes-21 to begin with, then on fertilization it becomes trisomic, and if the pregnancy goes to completion, then a child with Down's syndrome will be the result. (As discussed in Chapter 6, it is now possible to detect the presence of Down's syndrome while a fetus is only a few months old.) If the egg had no chromosome-21 to begin with, it has only one copy after fertilization; this apparently leads to early spontaneous abortion of the developing fetus because no living person who has a single copy of chromosome-21 has been reported. The result is the same if a normal egg is fertilized by an abnormal sperm produced by nondisjunction in the male. Also, nondisjunction can occur in the first few cell divisions following fertilization of a normal egg by a normal sperm. In this case, the cells with a single chromosome-21 die off, and development of the embryo proceeds by further division of those cells trisomic for chromosome-21.

1–19   Nondisjunction, the failure of chromosomes to sort out properly during the formation of sex cells, can result in Down's syndrome. When an egg with an extra chromosome is fertilized, the resulting zygote has three copies of chromosome-21.

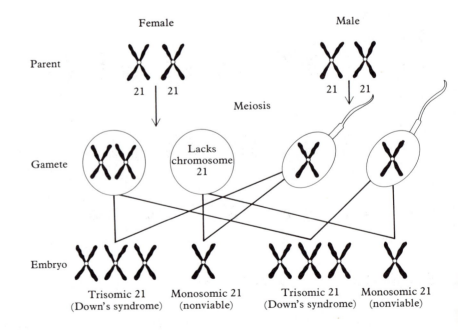

As we shall see, variations in the number of human sex chromosomes generally produce abnormalities that are less severe than those produced by the presence of abnormal numbers of autosomes. Trisomy for chromosomes-8, -9, -13, and -18 is occasionally observed among newborns, but those in whom it occurs usually die as infants or during the first few years of life. Examination of the chromosomes of fetuses that have been spontaneously aborted early in the pregnancy reveals that trisomy for most autosomes does occur, but it is usually not compatible with full-term development of the fetus (Table 1-2). Overall, it is estimated that at least 50 percent of all spontaneous abortions are associated with abnormal karyotypes; the number of living newborns with chromosomal abnormalities is about one-tenth the total number of fetuses that have abnormal chromosomes. On the other hand, deletions of an entire chromosome known as *monosomy*, is almost never observed in autosomes. Moreover, deletion of even part of an autosome often produces serious or fatal abnormalities. In the first reported example (1963) of the deletion of part of a human chromosome, the deleterious effects resulted from the deletion of one of the short arms of chromosome-5. Affected individuals have low birth weights and are mentally retarded, and their crying has a striking resemblance to that of the domestic cat, which led the French pediatricians who first described the condition to name it the *cri du chat* ("cat cry") *syndrome*.

The fact that trisomy-21 is compatible with life no doubt relates at least in part to the small size of the chromosome present in triplicate. Down's syndrome occurs with surprising frequency. It is observed in one out of 500 or 600 births and has been detected in up to one in 40 fetuses aborted before 20 weeks of gestation, and the chances for the occurrence of trisomy-21 increase with the age of the mother (Figure 1-20). Why is this so?

TABLE 1-2   FREQUENCY OF TRISOMY FOR AUTOSOMES IN 183 FETUSES THAT UNDERWENT SPONTANEOUS ABORTION.

Chromosome	Estimated frequency (%)
1	—
2	4.48
3	1.12
4	1.90
5	—
6	0.53
7	1.60
8	3.72
9	3.72
10	2.13
11	—
12	—
13	2.36
14	6.50
15	10.04
16	32.11
17	—
18	5.58
19	—
20	1.90
21	12.54
22	9.79
Total	99.99

SOURCE: From "A Cytogenetic Study of Human Spontaneous Abortions Using Banding Techniques," by M. R. Creasy, J. A. Crolla, and E. D. Alberman. *Human Genetics* 31(1976): 177–196.

1-20   Left, the age distribution of mothers of children who have Down's syndrome compared to that of all mothers. Right, data on 1119 cases of Down's syndrome in Victoria, Australia. (Left, after Dr. Victor McKusick; right, after Collman and Stoller, *American Journal of Public Health*, 52, 1962.)

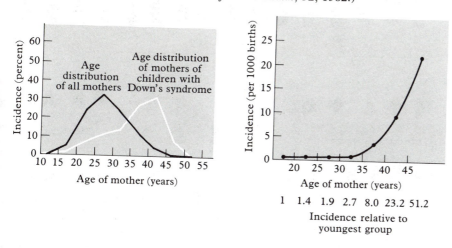

First, it is generally believed that the nondisjunction that leads to trisomy-21 occurs most often in the egg and not in the sperm cell. Perhaps this is because egg cells can sit in the ovary for decades before being ovulated and thus undergo some kind of metabolic or physical damage that later leads to nondisjunction during the production of sex cells. At any rate, the incidence of Down's syndrome increases with maternal age even if the age of the fathers is constant. In addition, Down's syndrome occurs with a rather characteristic frequency among the offspring of women of the same age, no matter how old their husbands are.

Although maternal age is generally associated with the incidence of Down's syndrome, certain circumstances greatly increase the chance that a particular pair of parents will produce an affected child at any age. These cases of Down's syndrome are not due to trisomy for chromosome-21, but rather are the result of an abnormality known as *translocation*, the fusion of two normal chromosomes, or at least parts of chromosomes. Most often the long arms of chromosomes-21 and -14 become fused to form a larger, composite chromosome known as *translocation chromosome*. A person who has one chromosome-21, one chromosome-14, and one translocation chromosome is phenotypically normal even though possessed of only 45 chromosomes (Figure 1-21). This is because all of the genetic material is represented and none is in excess. Such a person is called a *carrier of the translocation chromosome*. (The loss of the short arms of chromosomes-21 and -14 apparently does not have much effect on a person's phenotype.)

1–21   The 45 chromosomes of a woman who is a translocation carrier of Down's syndrome.

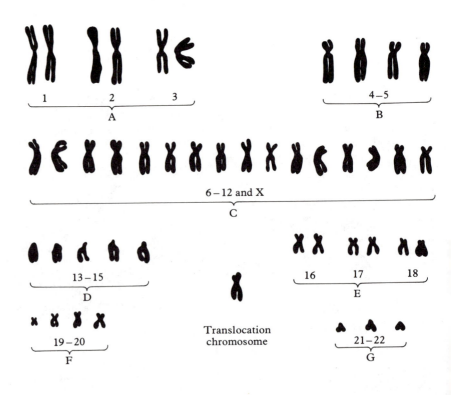

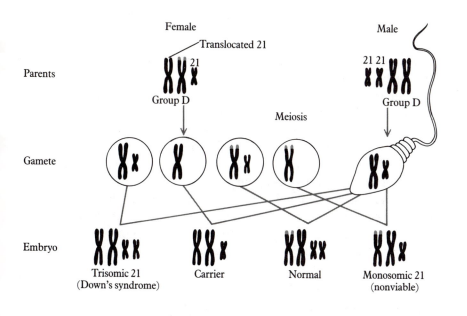

1-22   The kinds of sex cells and offspring that can be produced by a translocation carrier of Down's syndrome.

Carriers of a translocation chromosome can produce sex cells of two types—those containing the translocation chromosome and those without it (Figure 1-22). Either sex cells without the translocation chromosome are normal or they die before or shortly after fertilization because they contain chromosome-21 in but a single copy. On the other hand, after they are fertilized, sex cells containing the translocation chromosome result in the chromosome combinations shown in Figure 1-22. Overall we would expect Down's syndrome to occur regularly in the offspring of people who are translocation carriers, and this is what happens. For this reason, if a child with Down's syndrome is born to relatively young parents, they are advised to have their chromosomes studied to determine whether either parent is a translocation carrier for Down's syndrome and is therefore likely to produce another affected child.

With this discussion of autosomes and their abnormalities as background, we are now prepared to take up the subjects of sex determination and sex chromosomes. We do so in the following chapter.

Summary

Species retain their identity in nature because they are isolated from one another behaviorally and genetically. Variation between and within species is due to differences in genetic programs, and variation within a species is not randomly distributed. Some traits show definite patterns in their inheritance, as Mendel discovered.

Mendel proposed that alternate traits are passed from parents to offspring as independent units that are now called genes. He found that the factors responsible for some traits are dominant whereas others are recessive and discovered that he could predict what would happen in crosses involving several traits.

The discovery of the behavior of chromosomes during cell division provided a physical basis for Mendel's patterns. Human beings usually have 46 chromosomes in 23 pairs. Of these, 22 pairs are autosomes and the remaining pair are sex chromosomes.

In human pedigrees, autosomal dominant and recessive traits are usually associated with diseases or abnormalities. Dominant traits vary more in expressivity than recessive ones. Recessive traits can be manifested in a lineage by means of consanguineous marriages. Many human traits are determined by more than one pair of genes, and the expression of most traits depends in part on an individual's environment.

Abnormalities in the number of autosomes usually produce severe phenotypic abnormalities. The most common of these is Down's syndrome, which occurs more frequently among the offspring of older mothers and which can result from either nondisjunction or translocation.

Suggested Readings

The first six references are for those who would like to learn more about genetics in general. Each of these books is highly recommended. Some are more difficult than others, but all should be understandable to those who can make it through the present volume.

Principles of Human Genetics, 3d ed., by Curt Stern. W. H. Freeman and Company, 1973. Particularly strong in the areas of pedigree analysis and transmission genetics.

Human Genetics, 2d ed., by Victor A. McKusick. Prentice-Hall, 1969. Solid overview of all aspects of human genetics, especially population genetics.

Genetics, Evolution, and Man, by W. F. Bodmer and L. L. Cavalli-Sforza. W. H. Freeman and Company, 1976. Emphasizes the evolutionary aspects of human genetics.

Heredity Evolution and Society, by I. Michael Lerner and William J. Libby. W. H. Freeman and Company, 1976. Stresses the roles of ethical, social, and political factors in human genetics.

An Introduction to Genetic Analysis, 2d ed., by David T. Suzuki, Anthony J. F. Griffith, and Richard C. Lewontin. W. H. Freeman and Company, 1981. Excellent overview of basic genetics with emphasis on molecular aspects.

Human Genetics, by F. Vogel and A. G. Motulsky. Springer-Verlag, 1979. Rather technical survey of human genetics from a medical viewpoint.

Mendelian Inheritance in Man. Catalogs of Autosomal Dominant, Autosomal Recessive, and X-Linked Phenotypes, 6th ed, by Victor A. McKusick. Johns Hopkins University Press, 1983. Gives brief descriptions of all of the known human abnormalities that follow simple Mendelian patterns of inheritance.

"Has Mendel's Work Been Rediscovered?" by R. A. Fisher. *Annals of Science*, Vol. 1, 1936. (Reprinted in *The Origin of Genetics*, C. Stern and E. R. Sherwood, editors. W. H. Freeman and Company, 1966.) Discusses the evidence that Mendel may have fudged the results of some of his crosses.

"The Health of Abraham Lincoln—How Sick Was He?—Or Was He Sick at All?" by Harold Holzer. *MD Magazine*, Feb. 1983. Provides a good summary of the evidence for and against the belief that Lincoln had Marfan's syndrome.

"Chromosomes and Disease," by A. G. Bearn and James L. German III. *Scientific American*, Nov. 1961, Offprint 150. How advances in the visualization of human chromosomes opened up a new frontier in the study of human heredity.

Chapter 2

Sex Determination and Sex-Linked Traits

In the shallow offshore waters of several continents live some rather drab-looking but remarkable snails known as slipper shells (genus *Crepidula*). What makes these animals remarkable is the way their sex is determined. To simplify, the sex of a slipper shell depends on where it happens to land when it settles down to become an adult.

All young slipper shells are males, and they propel themselves through the water by means of winglike structures, as shown in Figure 2-1. However, when they lose their wings and settle to the bottom to begin their adult lives, the young males are transformed into females. That is, the young males are transformed into females if they do not land on a female. Young male slipper shells that land on females remain males, unless they become detached. If a mature male slipper shell becomes detached from a female, the male automatically changes into a female, as it would have done had it not landed on a female in the first place.

What does all of this activity accomplish? In brief, this unusual method of sex determination makes it likely that male and female slipper shells will live in the same area and will therefore mate with each other.

The slipper shell's method of sex determination is intriguing because it is so unusual. The sex of most animals, especially those with which we are more familiar, is determined at the time of fertilization and remains the same throughout life. Unlike slipper shells, most animals cannot adjust their sex according to circumstances. Sexual reproduction among animals that have separate sexes usually depends on individuals of the opposite sex finding and mating with each other. And under such circumstances, it is usually advantageous for a species to have about as many males as females.

Most of us would probably agree that the human species has roughly equal numbers of males and females, as do most other species that reproduce sexually. But the biological mechanism underlying the maintenance of this familiar, nearly equal distribution of the sexes eluded biologists until early in this century. At that time, investigators first turned their attention to male–female differences in chromosomes.

In 1902 it was discovered that the body cells of female grasshoppers contain one more chromosome than those of males. Shortly thereafter, the female's extra chromosome was rather romantically named the X *chromosome* (X for unknown), and it was suggested that the presence or absence of the X chromosome determined whether a grasshopper was a female or a male. Since then, it has been learned that chromosomes do have a role in sex determination for the great majority of organisms that have separate sexes, including most animals. But it turns out that male and female grasshoppers (and some of

Seventeen of the people in this 1894 photo are descendants of Queen Victoria (seated, center). She and the other two women indicated by an asterisk were carriers of hemophilia, a disease that is inherited as an X-linked recessive trait and that is characterized by a prolonged bleeding time. The other two women are Princess Henry (Irene) of Prussia (right) and Princess Alix (Alexandra) of Hesse (left), who later married Nicholas II, the last tsar of Russia. The future tsar is standing beside Alexandra. A pedigree of hemophilia in the royal families of Europe is found later in this chapter. (Photograph courtesy of the Gernsheim Collection, Humanities Research Center, University of Texas, Austin.)

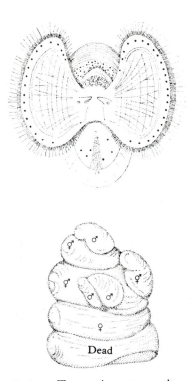

2–1   Top, an immature male slipper shell that has not yet shed his wings as seen from below. Bottom, a cluster of slipper shells. The male at the top of the stack is about the same size as the immature male. (♀ = female, ♂ = male, ♀̣ = an individual of intermediate sex in the process of changing from male to female.)

their close relatives)—with their unequal number of chromosomes—are the exception rather than the rule.

Male and female animals of the same species generally have the same number of chromosome pairs. But although all of the pairs are matched in females, there is one unmatched pair in males. As discussed in Chapter 1, the chromosome pairs that match in both sexes are called *autosomes*, and the members of the remaining pair, which match in females but not in males, are called *sex chromosomes*. For most animals, including all mammals, females are said to be of sex chromosome constitution *XX* and males of *XY*. (In birds the sex chromosomes match in males but not in females.)

Figure 2-2 compares the chromosomes of normal males and females. Notice that the X and Y chromosomes are easily distinguished from one another, just as, generally, are men and women. This chapter will discuss the chromosomal basis of maleness and femaleness in human beings and will then consider the patterns of inheritance and the special properties of traits determined by genes located on sex chromosomes. The first discussion concerns the observed human sex ratio and how it relates to the XY mechanism of sex determination.

The Human Sex Ratio

The XY chromosome mechanism of sex determination yields a reliable and nearly equal distribution of the sexes because of the sorting out of chromosomes during the formation of sex cells. Sex cells are produced by a form of cell division known as *meiosis*, which will be discussed in Chapter 5. For

2–2   The chromosomes of normal males and females. Left, males have 22 pairs of autosomes, one X chromosome, and one Y chromosome. Right, females have 22 pairs of autosomes and two X chromosomes.

now, it is enough to know that meiosis occurs only during the production of sex cells and that its overall effect is to reduce the number of chromosomes by half. Each sperm or egg receives one member of each pair of autosomes and either the X or Y sex chromosome. Thus, all normal eggs contain 22 unpaired autosomes and an X chromosome. But sperm do not. The body cells of males contain 22 pairs of autosomes, one X chromosome, and one Y chromosome. Thus, when men produce sex cells, meiosis results in two kinds of sperm with regard to sex chromosomes. All human sperm normally contain 22 unpaired autosomes, and on the average, half of them have an X chromosome and half have a Y chromosome. Thus, if X-bearing and Y-bearing sperm fertilize normal eggs and result in normal development about equally often, then we would expect the sexes to be about equally distributed, as shown in Figure 2-3. But are they?

Data concerning the number of males and females born in recent decades throughout the world reveal some rather surprising facts. By convention, the sex ratio is usually reported as the number of males per 100 females. Among American whites, the data show that approximately 106 boys are born for every 100 girls, so the sex ratio at birth is 106. This ratio varies somewhat from country to country and from one racial group to another. For example, the ratio is 113 in Korea, whereas it is about 102.6 among American blacks. Nonetheless, the worldwide data show that on the average, more males than females are born in every time interval for which reliable data exist.

The sex ratio at the time of fertilization, known as the *primary sex ratio*, is not necessarily the same as the ratio at the time of birth, the *secondary sex ratio*. Human males have at least a slight numerical edge on females at the time of birth. Is this because more females than males die as embryos? Apparently not; statistical studies on fetuses have revealed that the primary sex ratio is even higher than the secondary—perhaps as high as 130. Of course, there are many sources of error in determining the sex of fetuses, and the task of assigning a definite sex to the youngest embryos is most difficult. The possibility remains that the observed primary sex ratio results from a large number of female deaths at *very* early stages of development, a period for which very little data exist. But overall, it appears likely that human males do have numerical advantage over human females both at the time of fertilization and at the time of birth. How can this rather unexpected observation be explained?

If there really are more males than females at the time of fertilization, then several explanations are possible. For example, it may be that Y-bearing sperm win over X-bearing sperm in the race to the waiting egg. It has been suggested that because the Y chromosome is smaller than the X chromosome, Y-bearing sperm are lighter and therefore able to swim faster than X-bearing sperm. But, swimming is not the primary means by which human sperm reach the egg. Muscular contractions and currents within the female reproductive tract are primarily responsible for transporting human sperm to the oviduct (Fallopian tube), where fertilization takes place. However, Y-bearing and X-bearing sperm can be distinguished under the microscope, and there is no doubt that Y-bearing sperm do indeed swim faster. Also, the survival of Y-bearing sperm is favored in an alkaline (basic) environment, which exists

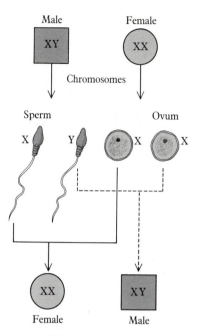

2-3 The sorting out of sex chromosomes during the formation of sex cells accounts for the nearly equal numbers of men and women.

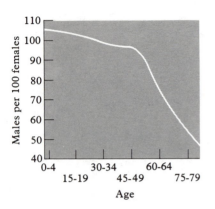

2–4   Data on the sex ratio in Scotland and Wales in 1960. The sex ratio of a given population for a given age is known as the *"tertiary" sex ratio.* (From A. S. Parkes.)

in the female lower reproductive tract at the time of ovulation. On the other hand, however, the survival of X-bearing sperm is favored in an acidic environment, which exists in the female reproductive tract about three days before ovulation. But the most important factor in determining the preponderance of male over female conceptions and births may simply be that Y-bearing sperm are more likely to survive than X-bearing ones. Several studies have revealed that perhaps twice as many Y-bearing sperm are present in the semen of normal men as X-bearing ones.

Human males start life with a numerical advantage over females, but they finish a weak second. This is because more males than females die at every stage of life from conception to old age. Thus, the numerical advantage of males at birth becomes progressively smaller until, at a certain age, the sexes exist in equal numbers. The exact age at which this occurs varies from population to population. As shown in Figure 2-4, in some countries males and females are equal in number by the time they reach 30 years of age; in others this happens as early as age 18 or as late as age 55. At whatever age it occurs, the numerical equality of males and females is not maintained. Females soon become the clear numerical majority.

It has been suggested that the male's greater susceptibility to death at every age may be related to the fact that men have only one X chromosome. (As we shall soon discuss, the Y chromosome is nearly a genetic blank for inherited human characteristics.) It is argued that males are more vulnerable to the effects of deleterious genes located on their single X chromosome. Even if women do have a deleterious gene on one X chromosome, they are likely to have a corresponding normal allele on their other X chromosome and thus are less likely to be severely affected. But arguments based on population genetics and recent advances in our understanding of the genetics of sex chromosomes have made it seem unlikely that men are at a disadvantage simply because they possess a single X chromosome. As we shall mention later in this chapter, in normal women only one X chromosome is genetically active in each cell; the other one is nonfunctional. Also, statistical studies of the habits of American men and women suggest that the relative longevity of women may be primarily determined not by genetic factors but by environmental ones. Specifically, fewer women than men smoke cigarettes. According to some researchers, the difference in life expectancy between American men and women completely disappears when deaths related to smoking are taken into account. But most researchers agree that, although differences in smoking habits may account for perhaps one-half of the difference in the average life span, genetic factors also play a role.

Voluntary, predictable changes in the primary and secondary sex ratios may soon be a reality. Technological means of altering the sex ratio of the human population already exist. For example, the secondary sex ratio could be altered by selective early abortion of fetuses of the unwanted sex; however, this means of controlling the sex of offspring is not likely to become widely accepted or widely available. A more appealing approach to voluntary sexual preselection of offspring is the separation of X- and Y-bearing sperm outside the body, followed by artificial insemination of either type of sperm. This technique has been successfully employed in animal husbandry and has been made more reliable by the recent development of a staining method that can identify the Y chromosome in living human cells, including sperm. Other less complicated techniques, such as the use of chemicals or prophylactics designed to block the entry of X- or Y-bearing sperm into the female reproductive tract, would probably be more widely accepted, but they are only theoretical possibilities at the present time.

What would happen to the sex ratio at birth if sexual selection were freely available in the United States? Of course, we can only speculate, but there is good reason to believe that the present ratio of about 106 male to 100 female births would probably not change significantly. Based on the data from sociological surveys, it has been predicted that what would change would be the probability that the first-born child would be a male and the second a female. What effect, if any, this change would have on our society remains to be seen, as do any other long-term effects of sexual preselection.

We now return to a discussion of the sex chromosomes themselves, and, in particular, to the relationship between femaleness and the X chromosome and maleness and the Y chromosome.

The Roles of the X and Y Chromosomes in Sex Determination—Abnormalities in the Number of Sex Chromosomes

What is the critical genetic difference between normal men and normal women? From our discussion so far, the answer would appear to be that women are of sex chromosome constitution XX and men are XY. But this tells us nothing of the exact relationship between the X and Y chromosomes and human maleness and femaleness. After all, the sexes are distinguished chromosomally not only by the male's having a Y chromosome, but also by his having only one X chromosome instead of two. Is it the presence of two X chromosomes that determines femaleness and the presence of a single X chromosome that results in maleness? Or does the Y chromosome determine maleness? These questions can now be answered. As so often happens in biology, our understanding of the normal process developed through the study of those persons in whom the normal mechanism of chromosomal sex determination had gone awry.

In 1949 it was discovered that two well-known but puzzling human afflictions, *Turner's syndrome* and *Klinefelter's syndrome*, are the result of abnormal sex-chromosome constitutions. Those who have Turner's syndrome are phenotypic females; that is, their genitalia are recognizably female. But these women are generally sterile because most of them have an underdeveloped uterus and no functional ovarian tissue. Other features of Turner's syndrome are short stature, distinctive facial features, a broad shieldlike chest, and a peculiar webbing of the neck (Figure 2-5). Those who are afflicted with Turner's syndrome have a single, unpaired X chromosome and are thus of sex chromosome constituion XO. *Sex chromosome constituion* can also be designated by giving the total number of chromosomes and indicating the sex chromosomes. In this notation, normal males are 46,XY, normal females are 46,XX, and persons with Turner's syndrome are 45,X.

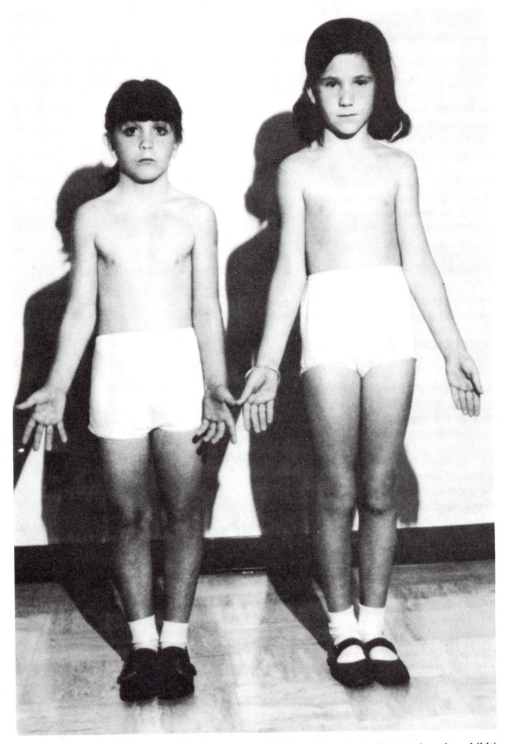

2–5   The 11-year-old girl on the left has Turner's syndrome; the other child is her normal 9-year-old sister. Note the affected girl's small stature, broad chest, and so-called webbed neck. Adult women with Turner's syndrome are sterile. (Photograph courtesy of Dr. Willard R. Centerwall, University of California, Davis School of Medicine.)

This discovery was made still more interesting when, in the same year, Klinefelter's syndrome was found to be associated with the sex chromosome constitution XXY (47,XXY). Persons who have Klinefelter's syndrome are phenotypic males, but they usually have very small testes and are sterile. Most of them are also very long-legged and have breast development resembling that of a mature woman.

How does an individual come to have an XO or XXY sex chromosome constitution? Usually by nondisjunction (a term you will recall from our discussion of Down's syndrome in the preceding chapter). Either the sex chromosomes fail to sort out properly during the production of sex cells by meiosis, or they fail to sort out properly in the first few cell divisions following fertilization (Figure 2-6).

The discovery of the chromosomal basis of Turner's and Klinefelter's syndromes was exciting, not only because the existence of the abnormal sex chromosomes could be explained as the result of nondisjunction, but also because of the insights it provided into the workings of the normal XY mechanism. The fact that those who have an XO sex chromosome constitution are phenotypic females, whereas those who have an XXY constitution are phenotypic males, suggests that the Y chromosome determines maleness and that those who have XO chromosomes appear female because, in the absence of the Y chromosome, even one X is enough to result in a female phenotype.

Klinefelter's syndrome occurs once in every 400 to 600 male births, whereas Turner's syndrome occurs once in about every 3500 female births. The reason for the comparatively low incidence of Turner's syndrome is that

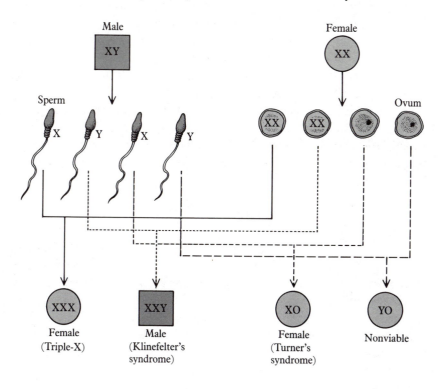

2–6   Nondisjunction can result in certain abnormalities in the number of sex chromosomes.

TABLE 2–1   HOW SEX CHROMOSOMES RELATE TO SEX PHENOTYPE. (BARR BODIES ARE DISCUSSED LATER IN THIS CHAPTER.)

Sex chromosome constitution	Sex phenotype	Number of Barr bodies
XX (normal woman)	Female	1
XY (normal man)	Male	0
XO (Turner's syndrome)	Female	0
XXY (Klinefelter's syndrome)	Male	1
XYY (see Chapter 8)	Male	0
XXX	Female	2
XXXY	Male	2
XXXX	Female	3
XXXXY	Male	3
XXXXX	Female	4
XXXXXY	Male	4

the vast majority of affected fetuses—perhaps as many as 98 percent—are spontaneously aborted. Since 1949, many cases of human abnormalities caused by unusual numbers of sex chromosomes have been reported. Most of these conditions occur less frequently than Turner's and Klinefelter's syndromes. Many of those who are affected are mentally retarded, and most have physical abnormalities. Sex chromosome constitutions of XXX, XXXX, and XXXXX have been reported, and those who have them are all phenotypic females. On the other hand, sex chromosome constitutions of XXXY, XXXXY, and XXXXXY are also known, and all belong to phenotypic males (Table 2-1). Thus it seems well established that in human beings the Y chromosome determines maleness and its absence results in a phenotypic female, as long as one or up to five X chromosomes are present. (The male-determining gene or genes on the Y chromosome are discussed in Chapter 5.)

Genetic Mosaics for Sex Chromosomes

Further evidence of the male-determining role of the human Y chromosome comes from the study of human *genetic mosaics*. These persons are remarkable in that their bodies consist of two or more cell lines (that is, cells with different numbers of chromosomes) side by side. Mosaics are the result of accidents that occur during cell division. Most often the accident consists of either nondisjunction or the accidental loss of a particular chromosome, and the error usually occurs in the first few cell divisions following fertilization. Mosaics *can* result from double fertilizations or from the fusion of two embryos very early in development, but this rarely happens.

Mosaics for sex chromosomes are encountered more frequently than those for autosomes (although the latter do occur). In fact, the first person with

GENETIC MOSAICS FOR SEX CHROMOSOMES 39

Klinefelter's syndrome whose chromosomes were studied was a sexual mosaic. Some cells in his bone marrow were XXY, whereas others were XX. This sex chromosome constitution is thus designated XX/XXY. Mosaicism is particularly prevalent in Turner's syndrome. More than one-third of all live-born infants with Turner's syndrome are mosaics with sex chromosome constitution XO/XX.

More recent studies have shown that mosaics in whom only X chromosomes are different (for example, XO/XX, XX/XXX, and XXX/XXXX, all of which have been reported) are phenotypic females (Table 2-2). Sexual mosaics in whom every cell has a Y chromosome, including XY/XXY, XY/XXXY, and XXXY/XXXXY, are phenotypic males. (Nonmosaic persons of genotype XYY are also phenotypic males. The XYY genotype is further discussed in Chapter 8.)

Of special interest are sexual mosaics whose bodies are made up of cells of different sex. As summarized in Table 2-2, such individuals are usually phenotypic males if one Y chromosome is present. Nonetheless, many of those who have both male and female cell lines exhibit some of the secondary sexual characteristics of the two sexes simultaneously. For example, XX/XXY mosaics may have one male and one female breast, bearded and unbearded facial areas, and, more important, testicular and ovarian tissue side by side in the same gonad (the sex cell-producing organ). Persons who have both kinds of sex tissue generally have a mixture of male and female features in their external genitalia and are known as *hermaphrodites*. (The word is derived from the names of the Greek deities Hermes and Aphrodite.)

Although most hermaphrodites are mosaics for cells of different sex, nonmosaic hermaphrodites of sex chromosome constitution XX (normal female) and XY (normal male) have also been reported. As you know, the human Y chromosome is strongly male determining. How then can we account for

TABLE 2–2   HUMAN SEX CHROMOSOMAL MOSAICS. THE MOSAICS MAY COMBINE TWO OR THREE CHROMOSOMAL CONSTITUTIONS. PHENOTYPICALLY, THE MOSAICS MAY BE FEMALE, MALE, OR MIXED.

Female	Male	Mixed
XO/XX	XY/XXY	XO/XY
XO/XXX	XY/XXXY	XO/XYY
XX/XXX	XXXY/XXXXY	XO/XXY
XXX/XXXX	XY/XXY/XXYY	XX/XY
XO/XX/XXX	XXXY/XXXXY/	XX/XXY
XX/XXX/XXXX	XXXXXY	XX/XXYY
		XO/XX/XY
		XO/XY/XXY
		XX/XXY/XXYYY

SOURCE: From *The Principles of Human Genetics*, 3d ed., by Curt Stern. W. H. Freeman and Company © 1973.

the presence of male characteristics in these rare XX individuals and female characteristics in these rare XY individuals?

The most widely accepted explanation is that the mixture of male and female features that exists in exceptional XX and XY individuals can arise in two ways: first, by the translocation of male-determining genes from the Y to the X chromosome, and second, by the influence of sex-determining genes located on the X chromosome and on autosomes. The translocation of male-determining genes to the X chromosome apparently accounts for a rare and baffling group of men: normal males who have fathered normal sons but who nonetheless are of sex chromosome constitution 46,XX. The influence of a sex-determining gene on an autosome or on the X chromosome (the exact location of the gene is not yet known) is well demonstrated by the *testicular feminization syndrome*, in which affected individuals are phenotypic females in spite of the fact that their sex chromosomes are XY. The existence of autosomal genes that play a role in sex determination has often been demonstrated in animals whose XY mechanism functions the same way as the human mechanism. There is no doubt that such genes are also important in human sex determination, but very little is known about them.

Most of the time, the sex-determining effects of autosomal genes are balanced by the XY mechanism. Thus, normal men and women have genes for maleness and femaleness, both on sex chromosomes and on autosomes. In men, male-determining genes on the Y chromosome and on autosomes outweigh the female-determining genes on the X chromosome and on autosomes. In women, female-determining genes on the X chromosome and on autosomes outweigh autosomal male-determining genes. (The latter are generally much weaker in their male-determining effects than is the presence of a Y chromosome.) But in rare instances, the effects of autosomal male-determining or female-determining genes may override the effects of sex-determining genes located on sex chromosomes, resulting in persons whose phenotypes and sex chromosomes are at odds with each other.

This concludes our discussion of human sex chromosomes as they relate to sex determination. We now turn to the patterns of heredity of genes that are located on the X and Y chromosomes. Such genes are said to be *sex linked* and have characteristic patterns of inheritance. The X chromosome probably has about as many genes as an autosome of similar length, but the Y chromosome, in spite of its strong male-determining effect is, as far as we know, almost completely lacking in other genes.

Y-Linked Inheritance

The pattern of inheritance of genes located on the Y chromosome is very simple. Only men have a Y chromosome, and all sons but no daughters receive this chromosome from their fathers. Thus, a man who manifests a trait that is determined by a gene on the Y chromosome will pass the trait on to all of his sons and to none of his daughters.

The restriction of a trait to males is not sufficient evidence to prove the existence of a Y-linked gene. This is because some autosomal traits are expressed only in males or only in females. But such *sex limited autosomal traits* can generally be distinguished from those determined by genes on the Y chromosome because traits that depend on autosomal genes are transmitted by both parents, whereas Y-linked traits neither appear in women nor are transmitted by them. On the other hand, *sex-influenced traits* are expressed by both sexes but have different patterns of heredity in males and females. The classic example is pattern baldness, which is an autosomal dominant trait in males and an autosomal recessive trait in females. Females require a double dose of the allele for pattern baldness to be expressed. This is probably due to the presence of larger amounts of the female sex hormone, estrogen, in females than in males.

Of the thousands of human traits known to have a genetic basis, only one is definitely known to be Y-linked. The trait is maleness. To be more specific, male-determining genes on the human Y chromosome somehow act on the undifferentiated gonad of the young embryo and cause it to differentiate into a testes. In the absence of the testicular differentiating gene, the gonad develops into an ovary. When the testicular differentiating gene is present, all of the body cells have on their surface a substance known as *H-Y antigen*. (An *antigen* is a foreign substance that stimulates the production of a specific antibody; H stands for "histocompatibility." The unique surface antigen of male body cells was discovered when it was observed that female mice reject skin grafts from males of the same inbred line. Histocompatibility antigens will be discussed in Chapter 7.) The presence of H-Y antigen on the cell surface is now considered to be such a reliable indicator of maleness that it can be used to ascertain the sex of infants whose sex is ambiguous at birth. It has recently been discovered that the testicular differentiating gene on the human Y chromosome has several copies, most of which are located on the short arm of the chromosome. The significance of multiple copies of the human male-determining gene is still unknown.

The only other trait that may be an example of Y linkage is a rather unromantic but harmless one known as hairy ear rims. Affected men have long, stiff hairs on the rims of their ears, as shown by three Muslim brothers from southern India in Figure 2-7. The trait also occurs among whites, Australian aborigines, and, more rarely, Japanese and Nigerian men. Not all instances of hairy ear rims can be attributed to the effects of Y-linked genes. In some cases, genes located on autosomes are clearly responsible for the trait. Nonetheless, in some groups, especially those from India, Y linkage appears to be a distinct possibility.

The human Y chromosome is not unique in its apparent genetic inertness (aside from its genes involved in sex determination). In some insects the Y chromosome is also known to be nearly devoid of genes other than those that determine sex. On the other hand, some fish have numerous Y-linked genes, as do some mice. In general, the Y chromosome remains poorly understood, but our knowledge of it will increase rapidly. Chromosomes are once again the objects of intensive study, as they were at the turn of the century. As we learn more about their structure, the role of the Y chromosome in sex determination and in Y-linked inheritance will become more clear.

Although the Y chromosome is nearly blank for inherited traits, the X chromosome is not. We now turn to the distinctive patterns of inheritance of traits determined by genes located on the X chromosome.

2–7   The strikingly hairy ear rims of three Muslim brothers from South India. This trait may be determined by a Y-linked gene. (Photograph by S. D. Sigamoni, Photograph Department, Christian Medical College Hospital, Vellore. From Stern, Centerwall, and Sarkar, *American Journal of Human Genetics, 16*. Copyright © 1964.)

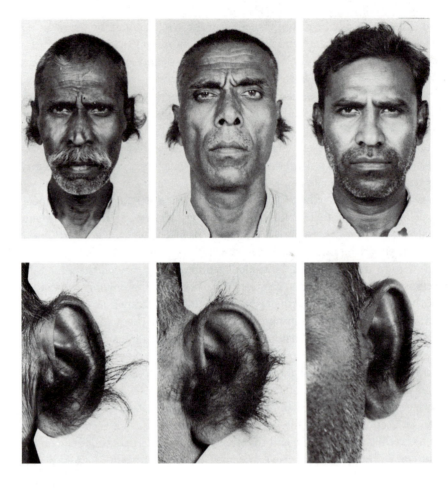

X-Linked Inheritance

About 120 abnormal traits are known to be determined by genes located on the X chromosome, and at least 140 more are suspected, but not proved, to be X-linked. (This means, of course, that the corresponding normal traits are determined by normal X-linked genes.) The X chromosome is by far the best-known human chromosome for two reasons. First, the patterns of inheritance of X-linked traits are very distinctive, and whenever we observe these patterns in a pedigree, we can usually conclude that the genes responsible for the traits are on the X chromosome. (As you know from our discussion of autosomal genes, pedigree analysis usually allows us to decide whether a trait is transmitted as an autosomal dominant or recessive, but it tells us nothing about the particular pair of autosomes on which the gene is located.) Second, assigning so many genes to the X chromosome allows us at least to begin to construct a genetic map of this chromosome—that is, to begin to pinpoint the exact location of the genes on the chromosome. As will be discussed in Chapter 5, the genetic map of the X chromosome is far better known than that of any other human chromosome.

Like autosomal traits, traits determined by X-linked genes may be either dominant or recessive. Nonetheless, X-linked traits have some peculiarities in their patterns of inheritance because of the presence of two X chromosomes in females and only one in males. Females, with their two X chromosomes, may be either heterozygous or homozygous for an abnormal X-linked allele. If the abnormal allele is dominant, then a woman who is heterozygous will manifest the trait. But this will not happen if the abnormal allele is recessive. In that case, a heterozygous woman usually does not manifest the trait, but instead is a carrier who appears normal.

On the other hand, males, with their single X chromosome, will always show the effects of an abnormal allele on their X chromosome regardless of whether the trait is inherited as a dominant or a recessive. In general, all males who have abnormal alleles on their X chromosomes manifest X-linked traits about equally. In other words, males who have X-linked traits are always affected; they cannot be unaffected carriers.

Of the 120 or so X-linked traits that are now known, only a few appear to be inherited as dominants. Included are brown discoloration of the teeth, the presence or absence of a particular antigen on the surface of red blood cells, and rickets resistant to vitamin D (a syndrome that is usually characterized by skeletal deformities and by low concentrations of phosphate in the blood).

What are the characteristic features of X-linked dominant inheritance? As shown in Figure 2-8, most women affected by X-linked dominant traits are heterozygous. (Recall that both men and women who manifest autosomal dominant traits are also usually heterozygous.) Women who are heterozygous for an X-linked dominant trait are affected, and they transmit the trait to both their sons and daughters. Affected men transmit the trait to all of their daughters, but to none of their sons. In a given pedigree, the fact that an affected father does not produce affected sons allows us to distinguish X-linked dominant traits from autosomal dominant ones. In autosomal dominant traits, both affected women and affected men pass the trait on to half of their daughters and to half of their sons. (Work this out for yourself.)

The great majority of human abnormalities known to be determined by genes on the X chromosome are inherited as recessive traits. As is also true

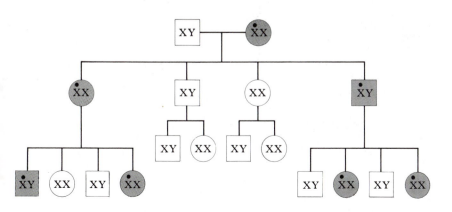

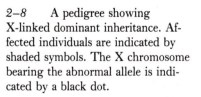

2–8   A pedigree showing X-linked dominant inheritance. Affected individuals are indicated by shaded symbols. The X chromosome bearing the abnormal allele is indicated by a black dot.

of X-linked dominant traits, a critical characteristic of X-linked recessive inheritance is the absence of father-to-son transmission. The following are some characteristics of X-linked recessive inheritance, all of which are summarized in the pedigrees shown in Figure 2-9.

First, virtually all persons affected by X-linked recessive traits are men. This is because X-linked traits are rare, and because in order for a woman to be affected, she must have an abnormal allele on each of her X chromosomes. In the absence of a spontaneous mutation, this could occur only if her mother was a carrier and her father was affected (Figure 2-9). Such rare, affected women have been recorded in human pedigrees. Second, all of the sons of affected men married to normal women are normal, whereas all of their daughters are carriers. Third, on the average, half of the sons of heterozygous (carrier) women married to normal men are normal, and half are affected (see Figure 2-9).

The most widely publicized human pedigrees are probably those in which the X-linked recessive trait known as *classic hemophilia* was transmitted throughout the royal families of Europe, particularly those of England and Russia. Persons who have hemophilia are sometimes called "bleeders." Their blood does not clot properly because of a deficiency of one of the many factors that participate in the normal clotting mechanism. (In classic hemophilia the deficiency is in factor VIII; other types of hemophilia result from deficiencies of different factors.) The affected persons are usually males (though rare female hemophiliacs have been reported), and they tend to bruise easily and to bleed heavily either into their joints, from their gums, or through tiny cuts, often as a result of relatively minor injuries.

2–9   Pedigrees showing X-linked recessive inheritance. Affected individuals are indicated by shaded symbols. The X chromosome bearing the abnormal allele is indicated by a black dot.

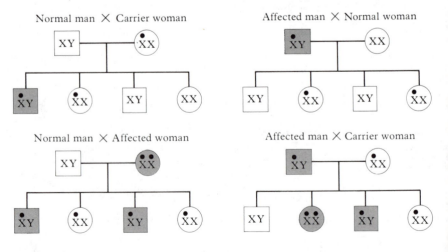

The pedigree of Queen Victoria and her descendants is shown in Figure 2-10. Analysis of it reveals that Queen Victoria must have been a carrier for classic hemophilia. One of her sons, Leopold, Duke of Albany, died of hemophilia at age 31. None of Victoria's forebears and neither her husband nor any of her then-living relatives had hemophilia, so the trait probably first appeared as a spontaneous mutation in one of the X chromosomes before Victoria inherited it from one of her parents. Alternatively, a mutation could have occurred in one of Victoria's X chromosomes during the queen's early embryonic life. Either way, the pedigree indicates that Victoria must have been a carrier. (The present royal family of England is completely free from the gene because Queen Elizabeth II traces her descent through Edward VII, one of Victoria's sons who did not have hemophilia.)

X-linked recessive genes are also responsible for some other familiar traits. Of these, perhaps the most common are certain kinds of baldness, red-green color blindness, and one form of muscular dystrophy, a disease in which the muscles of young males waste away in spite of the presence of an apparently normal nervous system.

For some traits determined by genes located on the X chromosome, the distinction between normal and abnormal is not always clear. One example is the trait known as *G6PD deficiency*, which results from the relative lack of the enzyme glucose-6-phosphate dehydrogenase, an enzyme that participates in carbohydrate metabolism. (Enzymes are discussed in the following chapter.) Affected persons are completely normal under most circumstances, but if they come into contact with certain environmental substances—ranging from the inhalation of the pollen of fava beans to the ingestion of primaquine, a drug used in treating malaria—there may be disastrous results. In the presence of these materials, among others, the red blood cells of affected individuals tend to break open; thus, severe anemia may result.

You will recall that heterozygous carriers of traits determined by genes located on autosomes can usually be identified by some kind of measurement, provided that the underlying biochemical defect is known. It turns out that persons who are heterozygous for autosomal traits usually have about half of the normal gene product for which they have an abnormal allele. For example, men and women who are carriers of an allele that results in the inability to digest milk sugar in homozygotes have about half of the normal concentration of the gene product that has a role in the digestion of this kind of sugar. This suggests that autosomal alleles contribute about equally to the total concentration of their normal biochemical product.

This raises an important question. Normal women have two X chromosomes, and normal men have only one. Therefore, to return to the example of G6PD, normal women have two normal alleles for the production of this enzyme, whereas men have only one. Is the concentration of the enzyme in the blood of men therefore only one-half of that in women? No, it is about the same in both sexes. And this is true not only of the enzyme G6PD, but of the biochemical products of X-linked genes in general. How can we account for this finding?

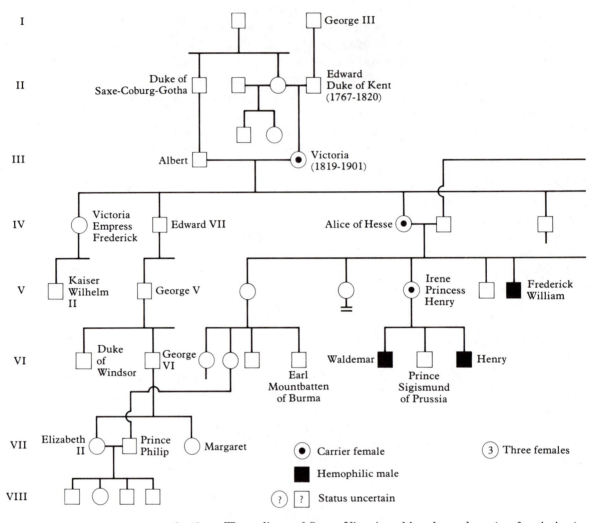

2–10   The pedigree of Queen Victoria and her descendants (see frontispiece) illustrates the transmission of the X-linked recessive trait, hemophilia. (After Dr. Victor A. McKusick.)

Dosage Compensation in X-Linked Genes— Lyons' Hypothesis

There are at least two possible explanations for the fact that women, with two X chromosomes, have about the same amount of any product determined by an X-linked gene as men, with their single X chromosome. First, the male's single X chromosome could work twice as hard—that is, produce twice the amount of gene product—as each of the female's X chromosomes. (This is true for the fruit fly.) Second, the activity of one or both of the female's X chromosomes could be less than that of the male's. There is little doubt that for the human species the second explanation is the correct one. Some of the earliest proof came from the study of persons who had abnormal sex chromosome constitutions.

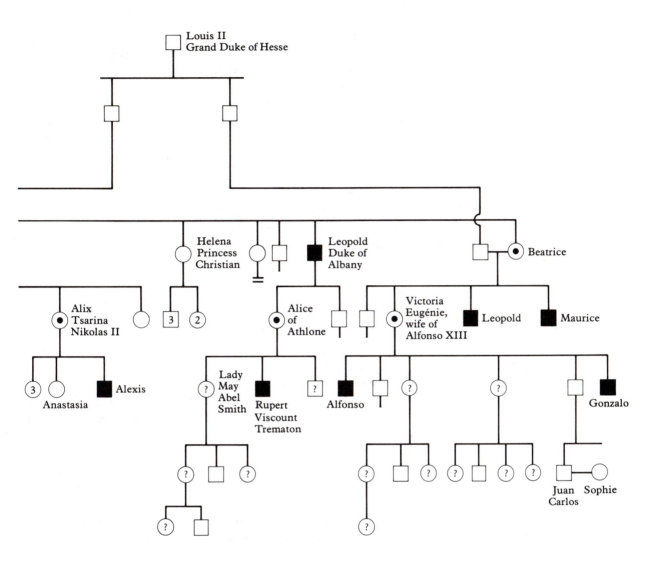

In the late 1940s it was discovered that the nondividing cells of female cats contain a small but distinct and stainable blob within their nuclei that is absent from the nuclei of male cats. This rather mysterious object became known as the *Barr body*, named after one of the first persons to describe it in detail. Barr bodies are found within the nuclei of most female cats from many kinds of animals, including humans (Figure 2-11). Nonetheless, Barr bodies are never observed within the nuclei of normal males.

Not long after Barr bodies were first discovered in the tissues of normal women, it was reported that women who had Turner's syndrome (XO) did *not* have a Barr body, whereas men who had Klinefelter's syndrome (XXY) *did*. This led to the hypothesis that the Barr body is actually an X chromosome that is tightly coiled in a dark-staining nuclear blob that is genetically inert. Thus, an obvious explanation for the occurrence of dosage compensation seemed to be this: women and men have the same concentration of gene products determined by X-linked genes because in normal women one X chromosome is genetically inactive.

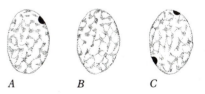

2–11   *A*, the nondividing nuclei of the cells of normal females contain a single Barr body. *B*, the nuclei of cells of normal males lack a Barr body. *C*, the nuclei of cells of persons whose sex chromosomes are XXX or XXXY have two Barr bodies. Also see Table 2-1.

Further support for the idea that the Barr body is an inactive X chromosome came from the study of the concentration of G6PD in persons who had abnormal sex chromosome constitutions. Thus, XO individuals (with no Barr body) and XXY individuals (with one Barr body) were both shown to have roughly normal concentrations of G6PD. Later, it was discovered that individuals whose sex chromosomes are XXX have normal G6PD levels and *two* Barr bodies. Similarly, persons who are XXXX have normal G6PD levels and *three* Barr bodies. The number of Barr bodies is thus one less than the total number of X chromosomes, which means that one X chromosome is always available to function normally.

The relationship between the inactivation of one X chromosome and dosage compensation is best summarized by *Lyons' hypothesis*, named after the person who was among the first to propose the idea. (The idea occurred almost simultaneously to several other investigators, and it has been refined over the years as more data have become available.)

The present version of Lyons' hypothesis is this. First, very early in embryonic life, when the number of cells in the body of a human female is relatively small, one of the X chromosomes becomes genetically inactive and forms a Barr body. In human embryos, Barr bodies are first observed at about five or six days after fertilization, and all of the cells of the early embryo apparently form them simultaneously. Second, in some cells, it is the mother's X chromosome that is turned off, and in others it is the father's. In other words, X chromosomes are turned off at random. Third, once the paternal or maternal X chromosome has been turned off in a given cell, the same X chromosome is turned off in all of the descendants of that cell during the later development of the embryo. We now turn to some of the evidence that one of the human female's X chromosomes really does behave in this remarkable way.

If one of the female's X chromosomes is turned off at random during early fetal development, then one would anticipate that normal women would be mosaics for the X chromosome, because different X chromosomes may be turned on in different cell lines. That this may be true is supported by studies showing that the red blood cells of women who are heterozygous carriers for G6PD deficiency clearly fall into two types. They have either normal enzyme activity or almost none. Presumably, the two different populations are the descendents of early embryonic cells in which opposite X chromosomes were inactivated.

Further evidence supporting the inactivation of one of the two X chromosomes in different female cell lines comes from determinations of the concentration of G6PD in the red cells of women who are affected by G6PD deficiency. As we would expect, most of these affected women are homozygous recessives; that is, they have an abnormal allele on each X chromosome. Nonetheless, some G6PD-deficient women are heterozygous, and although the concentration of G6PD within their red blood cells is usually intermediate, it may vary from as low as that of homozygous recessive women to as high as that of normal men and women. This makes sense if we assume that the red blood cells of heterozygous women with very low concentrations of G6PD are the descendants of embryonic cells in which the X chromosome that had the abnormal allele was the active one. Similarly, the red cells of heterozygous women who have very high concentrations of G6PD are presumably the descendants of embryonic cells that had the X chromosome whose abnormal allele was turned off. (Women have manifested X-linked recessive traits in spite of the fact that they are heterozygous for hemophilia and other traits. Such women are known as *manifesting heterozygotes;* presumably, most of their normal X chromosomes are inactivated by chance in early embryonic life.)

Although the evidence in favor of Lyons' hypothesis is generally convincing, there are some unanswered questions. Perhaps the most pressing one is this: Normal XX females and normal XY males both have only one active X chromosome. Why then is it that those who have Turner's syndrome (XO) and Klinefelter's syndrome (XXY), and who also have only one active X chromosome, are not only sterile but distinctly abnormal in several other ways?

It has been suggested that the abnormal phenotypes of XO and XXY individuals may result because inactivity affects most, but not all, of the X chromosome. If the portion of the mostly turned off X chromosome that supposedly remains active carries genes that determine the phenotypic differences between XO and XX individuals, then the phenotypic abnormalities of the XO genotype could be accounted for. Although XO individuals have one complete X chromosome, they lack the supposedly active portion of a normal woman's other, mostly inactivated, X chromosome. Similarly, individuals of genotype XXY would be abnormal because they have a normal X and a normal Y chromosome *plus* the active portion of the other, mostly inactive, X chromosome.

Recent discoveries strengthen this hypothesis: at least three genes, presumably clustered at the tip of one of the arms of the predominantly inactive X chromosome, are active in normal women. Therefore, it may be that products of these three genes (among others) are responsible for the wide range of physical abnormalities found in Turner's and Klinefelter's syndromes. However, the true explanation for the presence of these abnormalities remains a mystery.

Summary

In most sexually reproducing animals, it is advantageous to have about as many males and females, and these equal numbers are usually maintained by male–female differences in chromosomes.

All normal human beings have 22 pairs of autosomes; in addition, normal women have two X chromosomes, whereas normal men have one X chromosome and one Y chromosome. The sex of an individual is determined at fertilization and depends on whether the egg is fertilized by an X-bearing or a Y-bearing sperm.

The sex ratio of the human population varies with age. More males than females are conceived and born, but at every stage of life males are more likely to die than females. Voluntary selection of the sex of offspring may soon be a reality for some populations.

The discovery of the chromosomal basis of Turner's syndrome (XO) and Klinefelter's syndrome (XXY) suggested a male-determining role for the human Y chromosome, and this hypothesis was borne out by studies of human sexual mosaics. In the absence of the Y chromosome, the phenotype is female as long as at least one X chromosome is present. Autosomal genes must also influence sex determination, and both sexes have genes for maleness and femaleness on their autosomes.

The Y chromosome is nearly devoid of genes not involved in sex determination, but the X chromosome is known to carry at least 120 genes other than those that determine sex. Most X-linked abnormalities are inherited as recessive traits. X-linked traits, such as classic hemophilia, have distinctive patterns of inheritance.

Dosage compensation occurs for X-linked traits because although women have two X chromosomes, they do not have twice the concentration of products of X-linked genes that men, with one X chromosome, have. Lyons' hypothesis suggests that one of the female's X chromosomes is randomly inactivated very early in development. Inactivated X chromosomes, which can be observed within the nuclei of female cells, are called Barr bodies. The number of Barr bodies is always one less than the number of X chromosomes, and in normal women opposite X chromosomes may be active in different cell lines. Normal women are therefore mosaics for traits determined by X-linked genes.

Suggested Readings

"Sex Differences in Cells," by Ursula Mittwoch. *Scientific American*, July 1963, Offprint 161. A review of the major chromosomal differences between men and women.

Genetic Mosaics and Other Essays, by Curt Stern. Harvard University Press, 1968. A short, rather technical work for those especially interested in genetic mosaics.

"Sex Preselection in the United States: Some Implications," by Charles F. Westoff and Ronald R. Rindfuss, *Science*, vol. 184, 10 May 1974. What would happen to the secondary sex ratio if sex preselection were freely available in the United States?

"Rigging the Odds to Preselect the Baby's Sex," by Margaret Krie. *Sexual Medicine Today*, Dec. 1977. Discusses the physiological basis of preselecting a baby's sex and provides data on the relative numbers of X-bearing and Y-bearing sperm in normal semen.

"The Y Chromosome and Primary Sexual Differentiation," by Renée Bernstein. *Journal of the American Medical Association*, vol. 245, no. 19, 15 May 1981. A technical but readable paper on the testes-determining genes on the human Y chromosome and the role of H-Y antigen in the determination of maleness.

"Can Smoking Explain the Ultimate Gender Gap?" by Constance Holden. *Science*, vol. 221., 9 Sept. 1983. Offers evidence that smoking accounts for a large part of the difference in the average life span of men and women.

"The Origins of Men with Two X Chromosomes," by Paul Burgoyne. *Nature*, vol. 307, 12 Jan. 1984. Describes recent evidence that the translocation of male-determining genes to the X chromosome results in males with sex chromosome constitution 46,XX.

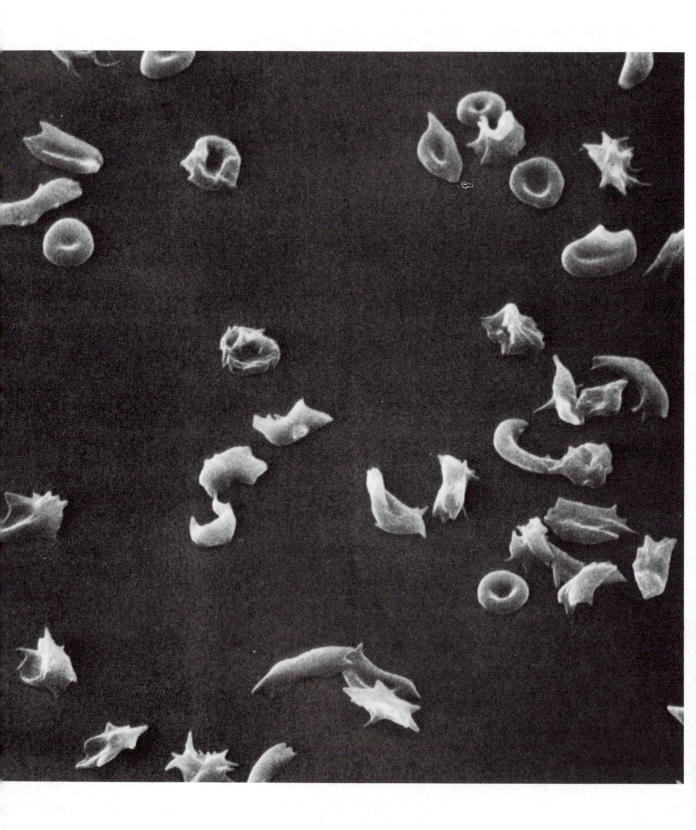

Chapter 3

DNA, Proteins, and Structural Genes

What is a gene, and what exactly does it do? Once it had been learned that genes, the units of heredity, are located on chromosomes, it was possible to approach these fundamental questions, to which nobody expected easy answers. By the mid-1920s, it was known that chromosomes are composed of only two kinds of biochemical substances, which are present in nearly equal amounts by weight. The first is protein, and the second is deoxyribonucleic acid, or DNA. A definite clue to which of these substances is the genetic program came in 1924, when it was discovered that all of the body cells of a given organism contain the same amount of DNA, whereas the sex cells contain half as much. This clearly suggested that DNA might be the hereditary material, but the idea was not widely accepted by biologists until the early 1950s. Before then, most biologists were convinced that the genetic material was not DNA, but protein, which is abundant in nondividing nuclei as well as in chromosomes. Proteins were known to be very complicated molecules, and what little was then known of the structure of DNA suggested that although its threadlike molecules were incredibly long, they were structurally simpler than protein molecules. Most biologists believed that protein was more likely to be the genetic program than DNA because complex protein molecules seemed more appropriate bearers of genetic information than the apparently simpler molecules of DNA.

The elucidation of the biochemical basis of heredity is one of the greatest intellectual achievements of this or any other century. In this chapter, we will consider the biochemical nature of the genetic program and of proteins. We shall also discuss the role of DNA in protein synthesis and how several abnormal human traits (including sickle-cell anemia and phenylketonuria [PKU]) can be explained in terms of biochemical alterations of the genetic program. Our present understanding of biochemical genetics and of how genes produce their effects is largely an outgrowth of a proposal of the detailed structure of DNA first made in 1953 by James D. Watson and Francis H. C. Crick. But before describing the structure and function of DNA, let us first examine how it was shown that DNA, not protein, is the hereditary material.

DNA Is the Genetic Program

In 1944, a series of experiments demonstrated that DNA alone, and not protein, could account for a phenomenon known as *transformation*, which had been discovered some years earlier. Transformation can be thought of as a lasting change in a cell's genetic program brought about by DNA from

Upon exposure to unusually low concentrations of oxygen, the red blood cells of persons who have sickle-cell trait may become distorted, as shown in this scanning electron micrograph. The donut-shaped object at the lower right is a normal red blood cell. The biochemical basis of sickle-cell trait is discussed in this chapter. (Photograph courtesy of Patricia Farnsworth.)

another cell. Thus, as shown in Figure 3-1, when purified DNA from dead bacteria of type I (which causes mice to die of pneumonia) is added to a suspension of living organisms of type II (which does not cause disease in mice), some living organisms that have characteristics of type I may form. Transformation occurs because some cells of type II take up type I DNA from the surrounding medium and thereby have their genetic programs altered; the chemical digestion of DNA, but not of protein, abolishes the suspension's ability to bring about transformation.

Once they are transformed, the organisms breed true to their new type, and it is possible to recover more type I DNA from the transformed bacteria than was originally added to the growing culture. Thus, not only does type

3–1   These experiments demonstrate that some kind of chemical substance produced by pneumonia-causing type I bacteria (smooth colonies) and present in an extract of heat-killed bacterial cells can transform harmless type II bacteria (rough colonies) into the deady type I variety. The substance responsible for transformation is DNA.

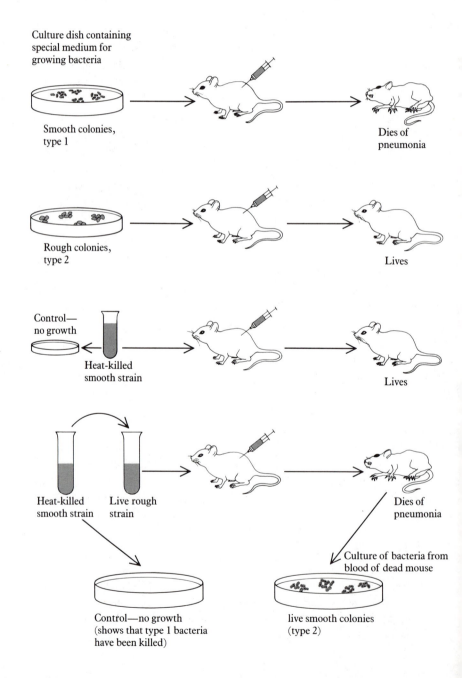

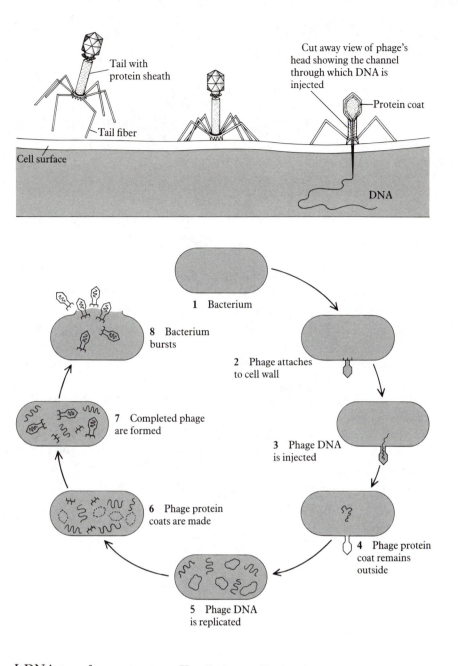

3–2 Top, a bacteriophage infects a bacterium by injecting DNA into the cell through the cell wall. The protein coat of the phage, which somewhat resembles a tiny lunar lander, does not enter the cell. Bottom, the injected DNA contains all of the information necessary to take over the cell and turn it into a phage-producing factory. The infected bacterium dies when it bursts and releases a new crop of infectious phage ready to seek out other bacteria.

I DNA transform some type II cells into cells that have some of the characteristics of Type I, but the resulting type I organisms reproduce and manufacture new type I DNA as they do so.

Further support for the idea that DNA is the genetic material came from the study of certain kinds of *viruses*. Viruses are molecular parasites, errant bits of genetic programs, that are made up of a protein coat surrounding a core of DNA. (In some viruses, the core is made up of a closely related substance, ribonucleic acid (RNA), discussed later in this chapter.) In 1952 it was reported that certain kinds of viruses that attack and usually destroy the bacterium *Escherichia coli* (*E. coli*), a normal inhabitant of the human digestive tract, are able to do so because they inject DNA into the bacterium like tiny parasitic syringes. Such a virus is known as a *bacteriophage,* or *phage* (see Figure 3-2). The phage's protein coat attaches the virus to the bacterium, but unlike the DNA, the protein coat does not get inside the cell. The DNA injected into the bacterium contains all of the information needed to manufacture entire new phage particles, including the protein coat.

Thus, by 1952 there was little doubt that DNA is the hereditary material. As we will discuss in later chapters, it is now known that the protein in the nondividing nucleus and in chromosomes provides a physical support, or scaffolding, for DNA molecules and plays a part in determining which genes are expressed and when. But it is in the enormously long, double-stranded, self-replicating molecule of DNA that we find the final, biochemical basis of heredity.

The Watson-Crick Model for the Structure of DNA

Like most other very large, naturally occurring molecules, DNA is made up of a few relatively simple chemical building blocks that are joined to each other in sequence by chemical bonds. In DNA, these building block compounds are called *nucleotides*. Each nucleotide is made up of three parts: a phosphate group, a sugar that contains five carbon atoms and is known as *deoxyribose*, and a nitrogen-containing base (Figure 3-3). (At this point, those of you who wish to review the meaning of the words *atom, molecule, chemical bond*, and others should consult Appendix II: "Some Simple Chemical Principles.")

There are four kinds of nucleotides in DNA. All of them contain the phosphate and the sugar, but their nitrogen-containing bases differ. The four bases fall into two categories according to their structure. Two of the bases, adenine and guanine, consist of double-ring, and the remaining two bases, cytosine and thymine, are made up of a single ring, as shown in Figure 3-4.

Within a DNA molecule, the nucleotides are bonded in such a way that the sugar of one nucleotide is always attached to the phosphate group of the

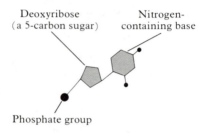

Deoxyribose (a 5-carbon sugar)   Nitrogen-containing base

Phosphate group

3–3   Schematic structure of a nucleotide, the building block of the DNA molecule.

3–4   The structural formulas of the four nitrogen-containing bases found in DNA. (Structural formulas are discussed in Appendix II.)

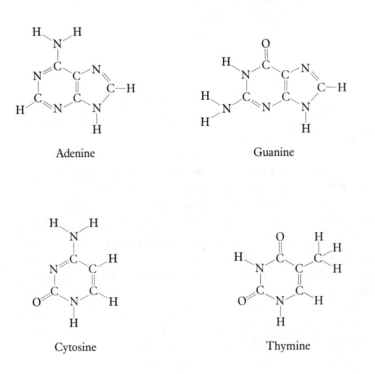

Adenine

Guanine

Cytosine

Thymine

next nucleotide in line. This arrangement results in a long chain consisting of alternating sugar and phosphate groups, and from this backbone the bases protrude (Figure 3-5). This was well known before Watson and Crick published their model for the structure of DNA. Their great accomplishment was not an explanation of the composition of DNA; rather, they described in detail the three-dimensional architecture of the molecule.

The model was based on information from several sources. It had been reported a few years earlier that all DNA molecules, regardless of their source, have something in common in the number of bases they possess. Although the amounts of the four bases vary widely from species to species, in all DNA molecules the number of adenine and thymine bases is exactly equal, and the number of guanine and cytosine bases is exactly equal.

Watson and Crick took these clues into account, along with the results of x-ray diffraction studies and other experiments aimed at determining the exact distances between atoms in the DNA molecule, and began to construct a physical model (first of cardboard and later of tin) that was consistent with all of the data (Figure 3-6). The publication of their model, in a report less than 1000 words long, in April 1953 was one of the most important events in the history of biology because it permitted an understanding of how genes function at the molecular level. The main features of the Watson-Crick model of the structure of DNA can be summarized as follows.

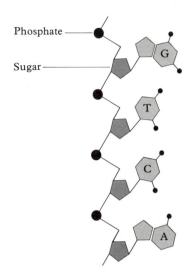

3–5 The backbone of the DNA molecule is made up of alternating sugar and phosphate groups. The nitrogen-containing bases point away from the backbone. Adenine (A) and guanine (G) are the double-ring bases; thymine (T) and cytosine (C) are the single-ring bases.

3–6 In their quest to determine the three-dimensional structure of DNA, J. D. Watson (left) and F. H. C. Crick (right) found it helpful to "play" with tin models of the four kinds of nucleotides that make up the larger molecule. In this way, the best fit between various combinations and orientations of the nucleotides was worked out.
(Courtesy of the Bettmann Archive.)

First, the DNA molecule is a double helix. To picture this helix, think of a backbone of alternating sugar and phosphate groups wrapped around a long, thin cylinder; this is one strand of the double helix. In the DNA helix, two strands are present, and they are held together by the bases, which protrude into the interior of the molecule and form weak chemical bonds with each other (Figure 3-7).

Second, the amount of adenine is equal to that of thymine because an adenine base and one strand is always bonded to a thymine base on the opposite strand. Similarly, the amount of guanine is equal to that of cytosine because in DNA these two bases are always bonded to each other across the double helix.

One of the most convincing points of this model was that it suggested how DNA might replicate itself. As Watson and Crick put it in their famous paper: "It has not escaped our notice that the specific pairing we have postulated immediately suggests a possible copying mechanism for the genetic material." Because adenine always pairs with thymine and guanine always pairs with

3–7   The structure of DNA. Top, a schematic, flattened view of the molecule. Adenine (A) pairs only with thymine (T), and guanine (G) pairs only with cytosine (C). The dotted lines represent weak chemical bonds known as *hydrogen bonds*, which hold the two strands of the double helix together. A and T are held together by two hydrogen bonds; G and C, by three. Bottom, three-dimensional representation of the double helix. Note that the two strands are intertwined and can be thought of as wound around an imaginary cylinder but running in opposite directions. (Top, from "The Synthesis of DNA," by Arthur Kornberg. Copyright © 1968 by Scientific American, Inc. All rights reserved. Bottom, from "The Repair of DNA," by Philip C. Hanawalt and Robert H. Haynes. Copyright © 1967 by Scientific American, Inc. All rights reserved.)

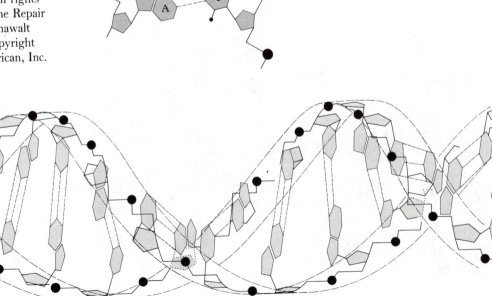

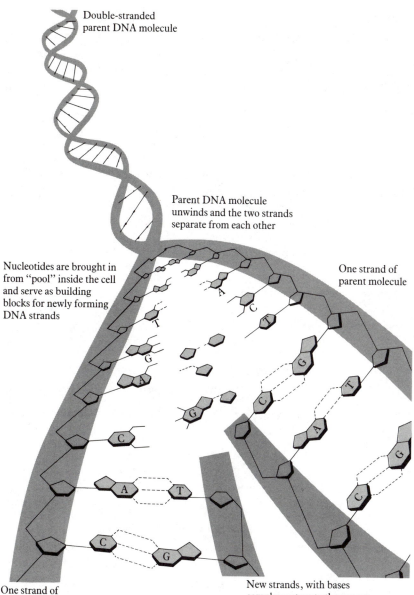

Double-stranded parent DNA molecule

Parent DNA molecule unwinds and the two strands separate from each other

Nucleotides are brought in from "pool" inside the cell and serve as building blocks for newly forming DNA strands

One strand of parent molecule

One strand of parent molecule

New strands, with bases complementary to the parent strand being assembled

3–8   During DNA replication the two strands of the double helix unwind and sepatate. Each strand then serves as a template for the manufacture of a new complementary strand. Nucleotides added to the newly replicated strands come from a large "pool" of nucleotides found inside the cell.

cytosine, if the two strands of DNA molecule are separated by breaking the bonds between the bases, each strand provides all the information necessary to synthesize a new partner. It was soon discovered that DNA is capable of self-replication, as any molecule reputed to be genetic material must be. As shown in Figure 3-8, DNA replicates by the separation of the two strands of the double helix; each strand then serves as a template for the manufacture of a new complementary strand.

Another reason that the Watson–Crick model created such excitement was that it suggested that genetic information might somehow be coded in the sequence of bases within the DNA molecule. At that time, no one knew how the double helix produced its biochemical effects. Nonetheless, within a few years, scientists in laboratories throughout the world began to contribute to a biochemical scheme of gene action whose details are still being worked out today. In sum, at the biochemical level, most genes participate in the process of protein synthesis, and within the double helix is encoded all of the information necessary for a cell to synthesize the proteins it needs to survive and reproduce.

What Is a Protein?

In 1838, the Dutch chemist G. J. Mulder coined the term "protein" for a class of compounds, which he was among the first to realize is, as he put it,

TABLE 3–1   A CLASSIFICATION OF PROTEINS BY FUNCTION.

Type of protein	Example
Structural	Collagen, the human body's most abundant protein, is found in various kinds of connective tissues that hold the body together.
Storage	Ovalbumin is a major source of materials and energy during early embryonic development.
Transport	Hemoglobin transports oxygen from areas of high concentration in the lungs to areas of lower oxygen concentration in the tissues.
Receptor	Insulin receptors are proteins found embedded in the cell membrane and exposed on the surface of the cell. When insulin and the receptor combine, sugar molecules enter the cell.
Hormone	Growth hormone, released by the pituitary gland, stimulates the growth of most body parts and has widespread metabolic effects.
Protective	Antibodies are produced in response to the presence of foreign substances, organisms, or tissues.
Contractile	Actin and myosin arranged in orderly arrays in muscle cells produce shortening by sliding past each other in a controlled manner.
Regulatory	Regulatory proteins that influence which genes are expressed and when are well known among bacteria and are also probably important in the expression of human genes.
Enzymes	Carbonic anhydrase markedly increases the rate at which CO_2 and H_2O combine to form H_2CO_3. Enzymes are the largest and most diverse class of proteins.

"without doubt the most important of the known components of living matter." The term comes from the Greek word *proteios*, meaning first, and it was well chosen. Proteins are the most complex and diverse of all known molecules. They are also the most abundant biochemicals within cells; at least 50 percent of the dry weight of all cells consists of proteins. Protein molecules are gigantic. A relatively simple one contains about 1000 atoms; the largest contain more than 1 million atoms. Each kind of cell and each species of living organism has many kinds of proteins that are unique to it. It has been estimated that there are as many different kinds of protein molecules on earth as there are stars in the Milky Way—roughly 100 billion.

Table 3-1 shows how proteins can be classified according to function. In this chapter, we are especially concerned with the transport protein, *hemoglobin*, and with the class of proteins known as enzymes. Hemoglobin is present in the blood of all vertebrates (fish, amphibians, reptiles, birds, and mammals) and is found only inside red blood cells. The function of hemoglobin is to combine with oxygen and to transport it from areas of high concentration (in the lungs) to areas of lower concentration (in the tissues), where the oxygen is released to participate in some of the myriad biochemical reactions that occur inside cells. (The total of all of the chemical reactions that take place inside living things is known as *metabolism*.) *Enzymes* are the most numerous of the major groups of proteins, and they account for most of the biochemical diversity among species and among individuals within a given species. *The function of enzymes is to speed up the rate at which specific biochemical reactions take place.* Without enzymes, the vast majority of the chemical reactions of living things could not take place fast enough to sustain them. Consider the simple chemical reaction in which a molecule of carbon dioxide (CO_2) combines with a molecule of water (H_2O) to form a molecule of carbonic acid (H_2CO_3), according to the equation $CO_2 + H_2O \rightleftharpoons H_2CO_3$. (The symbol $\rightleftharpoons$ indicates that the reaction can proceed equally well in either direction.) This reaction generally takes place inside red blood cells that contain an enzyme known as *carbonic anhydrase*, in whose presence the rate of formation of H_2CO_3 from CO_2 and H_2O is markedly increased. In fact, the presence of carbonic anhydrase, the reaction proceeds an amazing 10 billion times faster than in its absence.

All proteins have two things in common. First, their function depends on their shape, on their precise three-dimensional architecture; even a very minor alteration in certain parts of the carbonic anhydrase molecule will make it not function. Second, all proteins are made up of "building blocks" called *amino acids*, of which there are 20 common kinds (Table 3-2). Proteins consist of long chains of amino acids linked in tandem by a chemical bond known as a *peptide linkage*. Relatively short chains of amino acids are known as *polypeptides*, and many proteins are made up of several kinds of polypeptide chains joined together by chemical bonds. The number of amino acids in a protein molecule may vary from about 50 to 50,000 or more. In general, *the shape of a protein molecule, and therefore its specific function, ultimately depends on the sequence of the amino acids of which it is composed.*

Figure 3-9 shows the amino acid sequence of the protein hormone insulin, whose main function is to transport the sugar glucose from the bloodstream into the interior of most kinds of cells. Insulin is a very small, compact protein.

TABLE 3–2

AMINO ACIDS AND THEIR ABBREVIATIONS.

Amino acid	Three-letter abbreviation
Alanine	Ala
Arginine	Arg
Asparagine	Asn
Aspartic acid	Asp
Asparagine or aspartic acid	Asx
Cysteine	Cys
Glutamine	Gln
Glutamic acid	Glu
Glutamine or glutamic acid	Glx
Glycine	Gly
Histidine	His
Isoleucine	Ile
Leucine	Leu
Lysine	Lys
Methionine	Met
Phenylalanine	Phe
Proline	Pro
Serine	Ser
Threonine	Thr
Tryptophan	Trp
Tyrosine	Tyr
Valine	Val

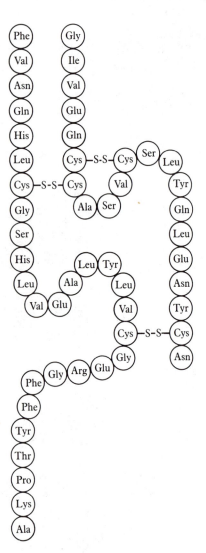

3–9   The insulin molecule consists of two short chains of amino acids joined by disulfide (S—S) linkages between cysteine molecules. An S—S linkage also occurs within the shorter chain.

The molecule consists of two relatively short polypeptide chains. The A chain is made up of 21 amino acids, and the B chain is made up of 30. The two chains are joined by chemical bonds that form between two molecules of the amino acid cysteine in various locations within the chains. The same kind of bond is also formed between two cysteine molecules within the A chain. All of these so-called disulfide linkages are important factors in determining the shape of the insulin molecule, on which its function depends. The shape of a protein also depends on the interactions between all of the other amino acids of which it is composed. These interactions, as well as the formation of the disulfide linkages, ultimately depend on the sequence in which the amino acids of each chain are joined together. And that is where DNA comes in. The link between DNA and proteins is this: *DNA contains coded information that specifies the order in which amino acids are joined together to form a particular protein.*

Protein Synthesis

The sequence of the bases in one of the strands of a particular segment of a DNA molecule determines the sequence of amino acids in a given protein. *Three* consecutive bases in DNA, known as a *codon*, code for *one* amino acid. For example, the base sequence TCA (thymine, cytosine, and adenine) codes for serine, whereas CAT (cytosine, adenine, thymine) codes for valine. Most amino acids have more than one codon; for example, GGC (guanine, guanine, cytosine) and GGA (guanine, guanine, adenine) both code for glycine.

Protein synthesis also requires a second kind of nucleic acid known as *ribonucleic acid* (*RNA*). In RNA, the five-carbon sugar is ribose, and RNA does not contain thymine but rather a different single-ring base, uracil, that pairs with adenine. Also, unlike DNA, most RNA is single-stranded. That is, the backbone of RNA is usually a single chain of alternating sugar and phosphate groups.

Protein synthesis begins when a segment of DNA within the nucleus unwinds and one of the strands provides the code for the synthesis of *messenger RNA* (*mRNA*). Messenger RNA is a single-stranded molecule with a sequence of bases that is complementary to the base sequence in a portion of one of the strands of DNA. Table 3-3 lists the three-letter code words for each of the amino acids in mRNA. The synthesis of the complementary mRNA from a DNA template is known as *transcription* (Figure 3-10).

In 1977, newly devised techniques of genetic engineering were first used to determine the exact sequence of bases in the portion of DNA that codes for one of the two kinds of polypeptide chains in hemoglobin, the *beta chain*. (The structure of the hemoglobin molecule is discussed later in this chapter.) The structure of this so-called beta-globin gene was a big surprise. The experiments revealed that the beta-globin gene is split into three regions that code for the amino acid sequence in the beta chain and demonstrated that the coding sequences are separated by two stretches of DNA base pairs that

TABLE 3–3   THE GENETIC CODE, AS IT APPEARS IN mRNA, WHICH IS COMPLEMENTARY TO ONE STRAND OF DNA. IN RNA, U REPLACES T AND PAIRS ONLY WITH A. (*ALSO SEE FIGURE 3–10.*)

First base in the codon	Second base in the codon				Third base in the codon
	U	C	A	G	
U	Phenylalanine	Serine	Tyrosine	Cysteine	U
	Phenylalanine	Serine	Tyrosine	Cysteine	C
	Leucine	Serine	*Termination*	*Termination*	A
	Leucine	Serine	*Termination*	Tryptophan	G
C	Leucine	Proline	Histidine	Arginine	U
	Leucine	Proline	Histidine	Arginine	C
	Leucine	Proline	Glutamine	Arginine	A
	Leucine	Proline	Glutamine	Arginine	G
A	Isoleucine	Threonine	Asparagine	Serine	U
	Isoleucine	Threonine	Asparagine	Serine	C
	Isoleucine	Threonine	Lysine	Arginine	A
	Methionine (*Initiation*)	Threonine	Lysine	Arginine	G
G	Valine	Alanine	Aspartic acid	Glycine	U
	Valine	Alanine	Aspartic acid	Glycine	C
	Valine	Alanine	Glutamic acid	Glycine	A
	Valine	Alanine	Glutamic acid	Glycine	G

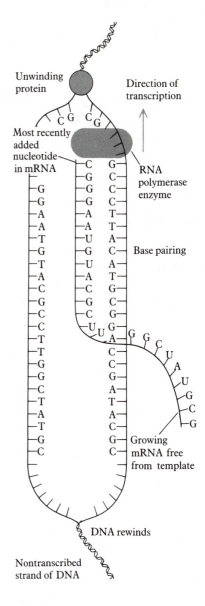

3–10   Schematic diagram of transcription—the synthesis of an mRNA with a base sequence complementary to a portion of one of the strands of DNA. [In RNA the base uracil (U) replaces the thymine (T) found in DNA, and U pairs only with adenine (A).] New bases are added to the growing mRNA by the enzyme RNA polymerase.

do not code for amino acids. As shown in Figure 3-11, one of the so-called intervening sequences, or *introns*, in the beta-globin gene is 550 base pairs long—longer than the longest coding sequence. Because the coding sequences are expressed (in that they do code for an amino acid sequence), they have been called *exons*. Further research has revealed that in the human genetic program, practically all regions of DNA that code for a specific protein include one or more noncoding introns; the noncoding stretches of base pairs are transcribed into an mRNA molecule along with the expressed sequences. Before the mRNA leaves the nucleus and begins to be translated into an amino acid sequence, the introns are removed by specific enzymes, and the expressed sequences are then spliced together to form an mRNA that contains all of the coding sequences but no introns. It is this so-called mature or *processed mRNA* that leaves the nucleus and directs the synthesis of a specific sequence of amino acids.

The *translation* of a mature mRNA into a specific amino acid sequence takes place on the surfaces of tiny particles known as *ribosomes*. Ribosomes

5′ [120 / 240] [500] [550] [250] 3′
Number of base pairs

ATGGTGCACCTGACTGATGCTGAGAAGGCTGCTGTCTCTTGCCTGTGGGGAAAGGTGAACTCCGATGAAGTTGC

| Met | Val | His | Leu | Thr | Asp | Ala | Glu | Lys | Ala | Ala | Val | Ser | Cys | Leu | Trp | Gly | Lys | Val | Asn | Ser | Asp | Glu | Val | Gl |

C/T/G CTAGGTGTACGTCGAACAGTGTCAGCTCGAGTGACTCCGACCGTTTCCACGGGAACTCCGACAGGTTCACTA

| Asp | Val | His | Leu | Lys | Asp | Cys | His | Leu | Glu | Ser | Leu | Ser | Ala | Phe | Thr | Gly | Lys | Leu | Ser | Asp | Leu | His | Asn |
Pro

A/G AACTTCAGG G/T/G

| Glu | Asn | Phe | Arg |

CTCCTGGGCAATATGATCGTGATTGTGCTGGGCCACCACCTTGGCAAGGA

| Leu | Leu | Gly | Asn | Met | Ile | Val | Ile | Val | Leu | Gly | His | His | Leu | Gly | Lys | As |

ACGGGTAGTCTGA

C/ACCCTTATACCTTCTTGGTAGTTGTATTGACATCTCGT TTTATGGTCT

| A |
| C |
| C |
| T |
| C |
| C |
| T |
| G |
| G |
| G |
| T |
| T |
| C |
| C |
| T |
| T |

C/CCTTGGCTATTCTGCTCAACCTTCCTATCAGAAAAAAAAGGGGAAGCGATTCTAGGGAGCAGTCTCCATGAC

3–11   Top, schematic drawing of the structure of the beta-globin gene, which is made up of three expressed regions or exons (dark shading) and two introns (light shading). Bottom, the complete sequence of nucleotides in the portion of a DNA strand that codes for the beta-globin gene of the mouse, as worked out by Philip Leder and his collaborators at the Harvard Medical School. The exons are darkly shaded; the introns are lightly shaded. The corresponding amino acid in beta-globin is indicated beneath each codon in the three expressed regions. (From "The Processing of RNA," by James E. Darnell, Jr. Copyright © 1983 by Scientific American, Inc. All rights reserved.)

are the sites where specially "activated" amino acids are strung together in the order coded by a specific mRNA. Amino acids that become incorporated into proteins are first activated by becoming attached to another kind of RNA known as *transfer RNA* (*tRNA*). Each amino acid has its own tRNA, and one part of each tRNA molecule has three exposed bases (known as an *anticodon*) that can pair with the complementary three-base sequence (codon) on the molecule of mRNA.

During protein synthesis, a mature mRNA becomes attached to a ribosome at one end, and *the ribosome then travels along the mRNA three base pairs at a time.* As the ribosome moves along the mRNA, each codon in the messenger is briefly exposed on the surface of the ribosome. When the codon is exposed, it is immediately "recognized" by the anticodon of a specific tRNA, with which it momentarily pairs. When this happens, the tRNA releases the activated amino acid, which is added to the end of the growing chain, and the tRNA is then released to pick up and activate another molecule of its specific amino acid (Figure 3-12). When the entire mRNA has been translated, it and the now completed amino acid chain are both released from the ribosome, which is then ready to begin to translate another mRNA.

Protein synthesis is an extraordinarily complex process; the description you have just read is a very simplified account of the myriad events that take place. This process demonstrates how the information in DNA is translated

CTTGGAGACAGAGGTCTGCTTTCCAGCAGACACTAACTTTC (A/G/T)

CATTGGACCTAT (G/G/T) TAGG (A/G/G) TTTTCCCTTTGTATCTGTCCCCTG (T)

GCCCTGGGCAGG (T) GCTGCTGGTTGTCTACCCTTGGACCCAGCGGTACTTTGATAGCTTTGGAGACCTA (T/C)

| Ala | Leu | Gly | Arg | | Leu | Leu | Val | Val | Tyr | Pro | Trp | Thr | Gln | Arg | Tyr | Phe | Asp | Ser | Phe | Gly | Asp | Leu/Ser |

AGCAATTTCCGTCAATAGTGGAAGAACGGGTACCCGGAAGTGAAACCGTAATGGGTACTATCGTCTCCGTCT

| Asp | Asn | Phe | Ala | Thr | Ile | Val | Lys | Lys | Gly | His | Ala | Lys | Val | Lys | Ala | Asn | Gly | Met | Ile | Ala | Ser | Ala | Ser |

CCCGCTGCACAGGCTGCCTTCCAGAAGGTGGTGGCTGGAGTGGCCACTGGCTTGGCACACAAGTACCACTAA

| Pro | Ala | Ala | Gln | Ala | Ala | Phe | Gln | Lys | Val | Val | Ala | Gly | Val | Ala | Thr | Ala | Leu | Ala | His | Lys | Tyr | His | Ter |

CCGAATAAAATGTTTCTGTAAACAAACGTTGGAATTTGATGACTGTTTAATAATATTCTTAGGATACAGTT (T/G/T/C)

TCTCAAGAATAGTAGCATAATTGGCTTTTATGCCAGGGTGACAGGGGAAGAATATATTTTACATATAAATT (C)

AGAAGGAACAGGAGACTCGTTCAATGTTCCGTTATAAATTTTTTCTGTTTGTTTGCATCTAGTTCTCTCTT (A/C/T)

TCCCCTCCTCCTTTCCTCCCAGTCCTTCTCTCTCTCCTCTCTCTCTTTCTCTAATCCTTTCCTTTCCCTCAG (T)

TTTATTGGTCGAATTAATTAAAATCATTTTTACGTTGACCCTTTAATTTTCGAATGGTTTCTTTCTCCTTTAC (T/G/T)

GAGTGTTGACAAGAGTTCGGATATTTTATTCTCTACTCAGAATTGCTGCTCCCCCTCACTCTGTTCTGTG (T)

into something concrete and useful, such as a molecule of hemoglobin or a specific enzyme. With this information as background, we are now ready to discuss the biochemical basis of some human abnormalities that result from defective proteins.

Structural Genes

The portion of one of the strands of a DNA molecule that codes for a specific protein or polypeptide chain is known as a *structural gene*. In the human genetic program, structural genes almost always consist of exons and introns, and proteins that are made up of more than one polypeptide chain may be encoded by more than one structural gene. The two polypeptides of insulin, for example, are coded for by a single structural gene, whereas the two kinds of polypeptides in hemoglobin are coded for by different structural genes. This section includes a description of three kinds of human abnormalities that result from defects in structural genes: sickle-cell anemia, beta thalassemia, and inborn errors of metabolism.

Sickle-Cell Anemia

The clinical syndrome known as *sickle-cell disease* was first described in 1910. At that time, little was known about it. First, those who were affected were

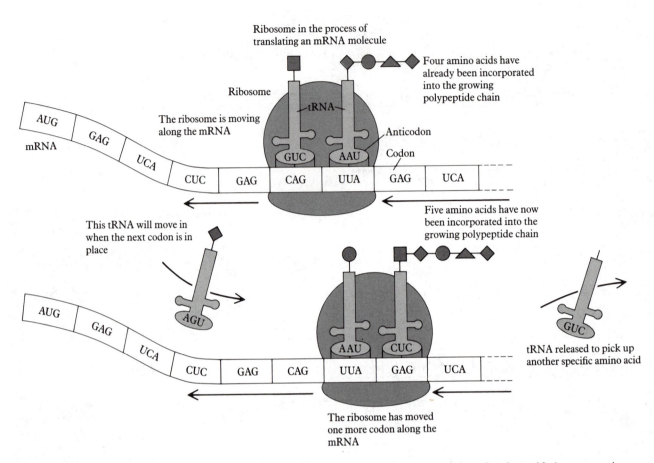

3–12   Schematic diagram of the incorporation of amino acids into a growing polypeptide chain that protrudes from the surface of a ribosome. An average human cell contains about 5 million ribosomes. The first codon in line is always AUG, which does not code for an amino acid but "recognizes" the surface of the ribosome, attaches the mRNA to it, and thus initiates translation.

blacks of both sexes, and many of them died of it in early childhood. Second, the disease was characterized by sometimes fatal crises that usually lasted for a few days. During a crisis, an affected person would develop fever and experience intense and incapacitating pain in the bones, large joints, abdomen, and elsewhere. Third, some of the red blood cells of affected persons were shaped like crescents, or sickles, and all of those who were affected had severe anemia. That is, the total number of red blood cells in their circulation was much less than normal.

A few years later, it was discovered that the sickling of the red cells is related to the state of oxygenation of the iron-containing pigment, hemoglobin. As you know, hemoglobin molecules inside red blood cells pick up oxygen in the lungs and release it to the tissues. Red blood cells from persons who have sickle-cell disease look normal when their hemoglobin molecules are saturated with oxygen. But when the saturation of hemoglobin by oxygen falls to lower than normal levels, sickling occurs; the process can be observed

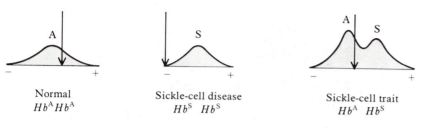

Normal
$Hb^A Hb^A$

Sickle-cell disease
$Hb^S Hb^S$

Sickle-cell trait
$Hb^A Hb^S$

3–13   In contrast to normal hemoglobin, the abnormal hemoglobin molecules of persons who have sickle-cell disease and sickle-cell trait (to be discussed later) show distinctive patterns of movement in an electric field. The areas under the curves correspond to the positions and amounts of hemoglobin molecules after an electric current has passed through the liquid in which the hemoglobins are dissolved. The arrow indicates the location of the specimen before the electric field was applied.

through the microscope. As the concentration of oxygen falls, more and more of the normal cells turn into sickle cells.

Why do the red blood cells of persons with sickle-cell disease change shape in the presence of low concentrations of oxygen? The first clue came in 1949, when it was discovered that the hemoglobin of such persons differs from normal hemoglobin. As shown in Figure 3-13, sickle-cell hemoglobin molecules were found to have a different pattern of movement than normal hemoglobin molecules in the presence of an electric field. This means that the two kinds of hemoglobin molecules cannot be identical. How are they different?

A refinement of the use of an electric field has provided the answer. Hemoglobin can be partially digested by enzymes that split the molecule into relatively short amino acid chains of varying length. If the resulting mixture of short amino acid chains is allowed to move under the influence of an electric field, a pattern, or "fingerprint," of the hemoglobin molecule is formed. Each spot of the fingerprint represents a different short chain of amino acids. When this is done for normal and sickle-cell hemoglobin, the patterns shown in Figure 3-14 result. Notice that protein digestion results in the formation of 26 short amino acid chains for both kinds of hemoglobin. In sickle-cell he-

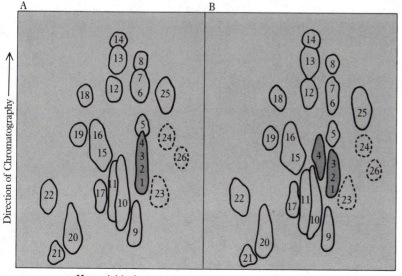

Hemoglobin A

Hemoglobin S

Direction of Chromatography

3–14   A "fingerprint" of normal and sickle-cell hemoglobins. Each spot on the fingerprint represents a short chain of amino acids.

3–15   Left, the hemoglobin molecule consists of four polypeptide chains, two alpha and two beta, plus four heme groups. The drawing at the bottom shows the three-dimensional shape of the beta chain. Each dot is the central carbon atom of one of the 146 amino acids. (From "The Hemoglobin Molecule," by M. F. Perutz. Copyright © 1964 by Scientific American, Inc. All rights reserved.)

Top view

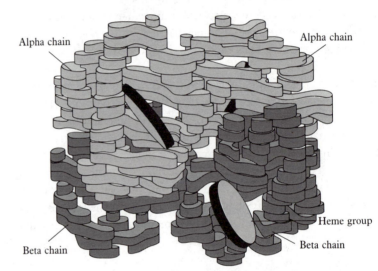

Side view

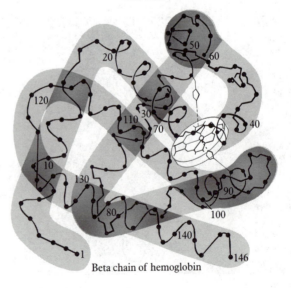

Beta chain of hemoglobin

moglobin, shown on the right in Figure 3-14, 25 of the 26 chains are identical to those in normal hemoglobin. Sickle-cell hemoglobin and normal hemoglobin differ in only one short amino acid chain, the one labeled "4" in Figure 3-14.

The exact nature of the difference between sickle-cell and normal hemoglobin is now known. As shown in Figure 3-15, the hemoglobin molecule contains two kind of polypeptides, the *alpha* chain and the *beta* chain. Each alpha chain consists of 141 amino acids, and each beta chain contains 146 amino acids. A normal hemoglobin molecule consists of two alpha chains, two beta chains and four nonprotein, iron-containing portions known as *heme*. As we have just discussed, when the hemoglobin molecule is partially digested by protein-splitting enzymes, 26 shorter amino acid chains are left and all but one of these shorter chains (chain 4) are identical in normal and sickle-cell hemoglobin. That chain on the fingerprint is a short chain consisting of eight amino acids that makes up one of the ends of the beta chain of hemoglobin. Of these eight amino acids, seven are identical in both kinds of hemoglobin. *Only one amino acid* out of a total of 287 is different in sickle-cell hemoglobin. Nonetheless, for persons who have sickle-cell disease, this seemingly trivial molecular variation can mean the difference between life and death.

As you may recall, three base pairs in DNA usually code for one amino acid in a protein molecule. In normal hemoglobin, the three-base sequence CTT (cytosine, thymine, thymine) codes for the amino acid glutamic acid. But in sickle-cell hemoglobin, the corresponding three-base sequence is CAT (cytosine, adenine, thymine), which codes for the amino acid valine. As shown in Figure 3-16, the amino acid substitution occurs sixth in line from one of the ends of the beta chain.

How does the presence of only one different amino acid out of 287 result in sickle-cell disease? It has recently been shown that the single amino acid difference has no effect on the stability of individual hemoglobin molecules; nor does it alter the molecule's oxygen-carrying ability. Rather, the single amino acid difference results in a unique reaction *between* individual molecules of sickle-cell hemoglobin. Deoxygenated molecules of sickle-cell hemoglobin spontaneously come together to form spiral, rigid, fiberlike structures that distort the red cell and result in sickling. (Most of the red cells quickly resume their normal shape when the hemoglobin is reoxygenated.) The rigid sickle cells tend to become trapped and broken when they circulate through capillaries, and the membrane of the red blood cell may be damaged in the sickling process. Both of these factors contribute to the breakdown of sickle cells that leads to the anemia that characterizes the disease.

At the present time, many of those who are affected by sickle-cell disease survive beyond the age of 50. But the improvement in outlook is not due to advances in our understanding of the exact biochemical nature of the disease. For all of our knowledge, we can do nothing to alter the composition of the DNA of those who are affected or to alter the translation of DNA into molecules of sickle-cell hemoglobin. The increased length of survival of affected individuals is largely the result of environmental factors such as better nu-

70

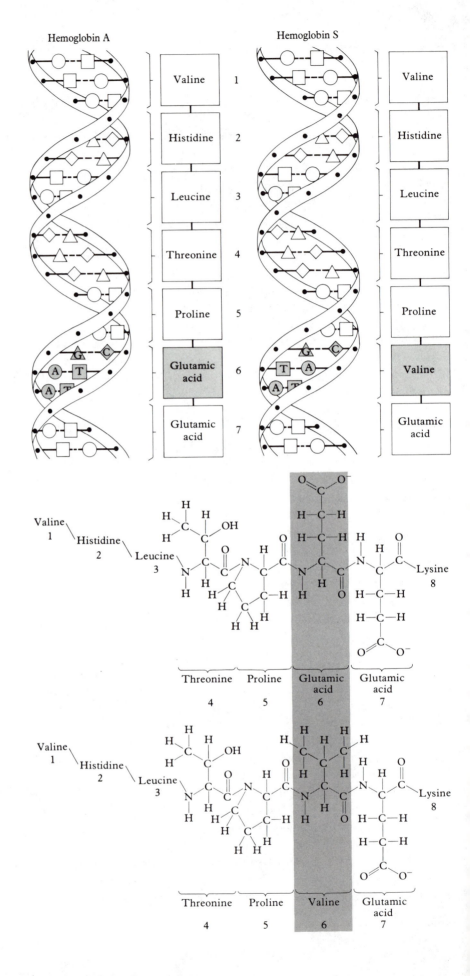

3–16 Only one amino acid out of a total of 287 differs in sickle-cell hemoglobin as compared to normal hemoglobin. As shown here, the amino acid valine replaces glutamic acid. The substitution occurs sixth in line from one of the ends of the beta chain. (From "The Genetics of the Human Population," by L. L. Cavalli-Sforza. Copyright © 1974 by Scientific American, Inc. All rights reserved.)

trition and the prevention of infections that may reduce the amount of oxygen available to the tissues and thus induce widespread sickling.

Shortly after sickle-cell disease was described in 1910, it was realized that the condition is transmitted as an autosomal recessive trait. This means that affected individuals are homozygous for the gene that results in the production of the abnormal beta chain in the hemoglobin molecule. The genotype of persons who have sickle-cell disease can be represented as $Hb^S Hb^S$. On the other hand, unaffected individuals are homozygous for normal hemoglobin, or hemoglobin A, and their genotype is $Hb^A Hb^A$. Persons who are heterozygous for sickle-cell hemoglobin are thus usually of genotype $Hb^A Hb^S$. Individuals of genotype $Hb^A Hb^S$ are said to have *sickle-cell trait* (not sickle-cell disease). Those who have sickle-cell trait do not have anemia and are perfectly normal under most circumstances. Nonetheless, some of the red cells of persons who have sickle-cell trait can be made to sickle in the laboratory by subjecting them to lower-than-normal concentrations of oxygen. As shown in Figure 3-13, persons who have sickle-cell trait can also be identified if samples of their hemoglobin molecules are allowed to move in the presence of an electric field. As we might expect, persons who have sickle-cell trait usually have approximately equal amounts of normal and sickle-cell hemoglobins throughout their red cells.

Sickle-cell trait occurs in about 8 percent of blacks in the United States, most of whom are of African ancestry. Sickle-cell disease, on the other hand, is becoming less common in the United States, but it still occurs. At least some of the decrease in the number of new cases reported per year is probably related to increased public awareness of the disease and to the detection of normal-appearing individuals who have sickle-cell trait (see Chapter 7). But black people are not the only human beings afflicted by sickle-cell disease. The abnormal gene also occurs relatively frequently in people inhabiting the Mediterranean area, Arabia, and India. Clearly, people who have sickle-cell disease are at a disadvantage compared to those who have normal hemoglobin. Why, then, is the gene that in the homozygous condition results in sickle-cell disease so widespread in different areas of the world, particularly Africa?

Figure 3-17 is a map showing the distribution of the sickle-cell gene in various parts of the Old World. The highest frequencies are in a rather broad belt across the African continent in which a severe form of malaria is a common cause of death. It has been found that the red blood cells of those who have sickle-cell trait ($Hb^A Hb^S$), which contain a mixture of normal and sickle-cell hemoglobin, are more resistant to malaria than normal red blood cells. This advantage probably accounts in large part for the persistent widespread distribution of a gene that in the homozygous state results in reduced reproductive fitness and may lead to early death.

Beta Thalassemia

In the past few decades, new research has discovered that hundreds of abnormal human hemoglobins cause anemias of varying severity, especially in individuals who are homozygous for a defective allele. One important group of anemias, known as *thalassemias*, is characterized by a decrease in the rate

3–17   The distribution of the
sickle-cell gene in various parts of
the Old World correlates with the
distribution of malaria, shown by
the shaded area in the map on the
facing page. (From "The Genetics of
the Human Population," by L. L.
Cavalli-Sforza. Copyright © 1974
by Scientific American, Inc. All
rights reserved.)

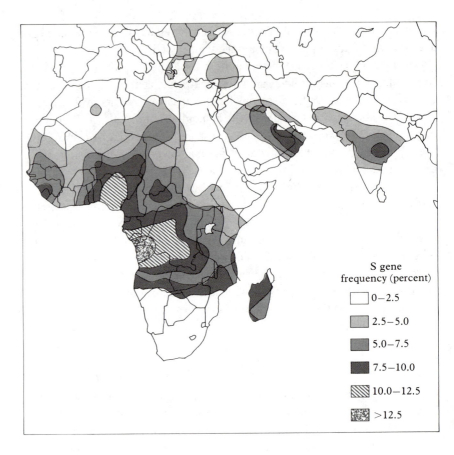

S gene
frequency (percent)

☐ 0—2.5

▨ 2.5—5.0

▨ 5.0—7.5

■ 7.5—10.0

▨ 10.0—12.5

▨ >12.5

of synthesis of the alpha or beta chains. The reduction in the rate of synthesis
depends on defects in the structural genes that code for either the alpha or
the beta chain.

Beta thalassemia results from a decrease in the rate of synthesis of beta
chains. Three different abnormal beta chains can result in clinical beta thalassemia. The severity of the disease is determined by the type of abnormal
beta chain and by whether the affected person is heterozygous or homozygous
for the defective allele. Severely affected homozygotes are said to have *thalassemia major*, which is associated with profound anemia, retardation of
growth, and increased susceptibility to infections, among other abnormalities.
On the other hand, heterozygotes have *thalassemia minor*, usually with mild
anemia that does not interfere with their daily activities.

One form of beta thalassemia is worthy of special mention. In 1982 it was
discovered that beta thalassemia can result not only from a defect in one of
the coding portions (exons) of the human beta-globin gene, but also from a
defect in one of the noncoding introns. Specifically, the abnormal gene has
a thymine where a guanine is normally found within the first intron, at a
point 19 bases away from the junction with the second exon (see Figure 3-

11). The presence of the "wrong" base at this particular location does not affect transcription, but it causes the mRNA for the beta chain to be improperly "edited" and "spliced" before it is exported from the nucleus to be translated on the surface of a ribosome. The improper splicing out of the intron and the subsequent incorrect splicing together of the exons have a disastrous effect on the translation of the abnormally processed mRNA. The first 29 of the normal 141 amino acids are strung together in the proper order, but the defective splicing so distorts the sequence of base pairs from that point on in the mRNA that the next 6 amino acids incorporated are incorrect. After a total of 35 amino acids has been added to the chain, the mRNA is so "garbled" from improper editing that the ribosome can no longer translate it, and the abnormal mRNA and the short, defective polypeptide are both released. The short, abnormal segment of the beta chain cannot combine with normal alpha chains to form hemoglobin; severe anemia then results in individuals who are homozygous recessive for the defective structural gene. This is the first reported example of a genetically determined human abnormality that results from the improper processing of an mRNA molecule.

Inborn Errors of Metabolism

In 1901, the British physician Archibald Garrod published an important study concerning a rare disease known as *alcaptonuria* (Figure 3-18). Persons who have this disease appear normal during childhood, but as adults they develop a blue-black discoloration of the ears, the whites of the eyes, the tip of the nose, and other areas of the body where cartilage lies just beneath the skin. Also, on exposure to several hours of sunlight, the urine of those who have alcaptonuria turns jet black. Both the black urine and the discoloration of cartilage result from the buildup of a substance called *homogentisic acid*. This compound is also deposited in the cartilage of large joints, and those who have alcaptonuria may therefore develop severe arthritis.

Garrod described 11 cases of this rare disorder and noted that at least 3 of the persons he studied were the offspring of parents who were closely related to each other; they were the offspring of consanguineous matings. Garrod used this observation, along with his understanding of Mendel's ratios (which had been rediscovered only the year before), as the basis for a bold and insightful explanation of the nature of the defect that causes alcaptonuria. Garrod suggested that the condition was due to the effects of a single recessive gene that results in the manufacture of a defective enzyme. The disease was thus what Garrod called an *inborn error of metabolism* and human abnormalities that result from defects in structural genes that code for enzymes are still so termed. However, the definition of inborn error of metabolism has recently expanded to include all diseases resulting from defects in structural genes. Inborn errors that result from defective enzymes are by far the largest group, reflecting the complexity of the intricate web of enzyme-dependent biochemical reactions that together constitute metabolism.

Garrod's explanation proved to be correct. The exact nature of the biochemical defect in alcaptonuria is now known. As just mentioned, affected persons have abnormally high concentrations of homogentisic acid in their urine and cartilage. This excess results because they have very low, or nonexistent, concentrations of an enzyme that is usually responsible for the further metabolism of homogentisic acid.

Homogentisic acid is one of the metabolites of the amino acid tyrosine. In normal individuals the following metabolic pathway operates:

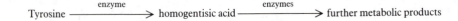

$$\text{Tyrosine} \xrightarrow{\text{enzyme}} \text{homogentisic acid} \xrightarrow{\text{enzymes}} \text{further metabolic products}$$

3–18   Sir Archibald Garrod coined the term *inborn errors of metabolism* early in the twentieth century. (Courtesy of The Royal Society.)

But in persons who have alcaptonuria, the lack of specific enzyme results in a metabolic block to the further processing of homogentisic acid, which therefore accumulates in their tissues, where it may produce its undesirable effects. This block can be represented as follows:

Tyrosine $\xrightarrow{\text{enzyme}}$ homogentisic acid $\xrightarrow{\text{enzymes}}$ further metabolic products

Garrod later showed that albinism, a condition discussed in Chapter 1, is also the result of a metabolic block that affects the amino acid tyrosine. Tyrosine is not only the precursor of homogentisic acid, but also the basic building block of the pigment melanin.

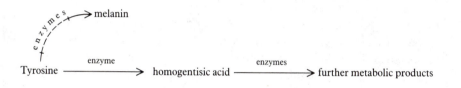

The most common kind of albinism results from the lack of an enzyme that participates in the manufacture of melanin from tyrosine. (But albinism can also result from a different inborn error whose overall effect is to greatly reduce the amount of tyrosine available to form melanin. In the second kind of albinism, affected persons have a normal concentration of the enzyme that facilitates the conversion of tyrosine to melanin.)

Since Garrod's time, we have become aware of yet another inborn error of metabolism in the metabolic pathway we have been discussing. This defect is known as *phenylketonuria (PKU)*. PKU is a serious disease, because if untreated, it can result in severe mental retardation. PKU results from a defect in the enzyme that converts the amino acid phenylalanine into tyrosine.

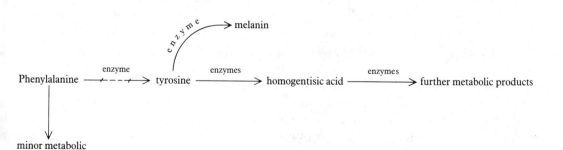

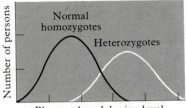

3–19   The phenylalanine tolerance test can identify some heterozygous carriers of PKU. After a large oral dose of phenylalanine is given, the level of the amino acid in the blood of carriers is higher and stays elevated longer than in normal individuals.

The lack of the appropriate enzyme for converting phenylalanine into tyrosine has two major effects. First, the concentration of phenylalanine in the tissues is greatly elevated. Second, minor metabolic products of phenylalanine that are normally present in only small quantities also accumulate. (It may be the buildup of these minor metabolic products during the first few months and years of life that accounts for the mental retardation.) Because they cannot convert phenylalanine into tyrosine, persons affected by PKU are lightly pigmented. Nonetheless, enough tyrosine is directly available in the diet so that they can still manufacture considerable quantities of melanin (and of homogentisic acid).

PKU can be effectively treated by greatly reducing the amount of phenylalanine present in the diet of an affected infant. Luckily for those who suffer from PKU, phenylalanine is one of the 10 essential amino acids that the human body cannot manufacture for itself; all of the body's phenylalanine comes directly from the diet. By restricting dietary intake, the buildup of high concentrations of phenylalanine and its minor metabolic products, and thus the mental retardation associated with the disease, can be prevented. At the same time, enough phenylalanine must be provided to allow for normal growth. Affected children are usually kept on their special diet indefinitely. If the diet is not restricted in the amount of phenylalanine, the school performance of these children deteriorates.

Like most other inborn errors of metabolism, PKU is inherited as a recessive trait and is manifested only by homozygotes. But it is possible to identify some carriers of the abnormal gene by subjecting them to the phenylalanine tolerance test. In this test, a large dose of phenylalanine is administered orally and the rate at which it disappears from the bloodstream is measured. As shown in Figure 3-19, when heterozygous carriers of PKU are fed a standard dose of phenylalanine, they show higher and more prolonged elevations of phenylalanine in their blood than do normal persons. This is because heterozygous carriers of PKU, although they appear normal, have a lower than normal concentration of the enzyme that aids in converting phenylalanine to tyrosine. As Figure 3-19 indicates, there is considerable overlap between the tolerance curves of carries of PKU and normal homozygotes. For this reason, the phenylalanine tolerance test has never been widely adopted. As will be discussed in Chapter 6, this limitation has recently been overcome by the technique of DNA analysis, which now allows the reliable detection of PKU carriers and of affected fetuses.

In recent years, the number of human diseases known to be due to inborn errors of metabolism has skyrocketed. More than 200 diseases that result from abnormally low concentrations of critical enzymes have already been described, and the list continues to grow. The metabolic pathways for phenylalanine and tyrosine just mentioned are but two examples of the thousands of pathways that interconnect chemical reactions inside cells. There is every reason to expect that eventually some kind of disease or phenotypic abnormality will be associated with a defect in every step of each metabolic pathway. Each disease would therefore result from a defective enzyme that catalyzed a specific step.

But the ability to *treat* diseases caused by inborn errors of metabolism has not generally kept pace with the ability both to diagnose them and to identify carriers by means of some kind of test. It is sometimes possible to prevent the development of undesired effects by limiting dietary intake, as with phenylalanine and PKU. But more often than not, we can do little or nothing to affect the course of diseases that result from inborn errors of metabolism. It is not now possible to treat these diseases simply by replacing the missing or defective enzymes. This is because enzymes generally do their work inside cells, and even if we are able to prepare concentrated extracts of the missing or defective enzyme, we have no way of getting them inside the cells that need them to speed up the rates of crucial biochemical reactions. A more fruitful approach may be the introduction of normal structural genes into the defective genetic programs of affected individuals. This experimental means of treating inborn errors of metabolism will be discussed in Chapter 6.

At present, and probably for some time to come, the best way to deal with diseases due to inborn errors of metabolism is to prevent them. There are two main ways of doing this. First, genetic counseling, sometimes in combination with tolerance tests, can help identify apparently normal carriers among the relatives of an affected person, or among other persons who might be carriers. Second, it is now possible to detect many inborn errors of metabolism in cells obtained from human embryos during the first few months of development. Thus, parents might choose to terminate a pregnancy voluntarily if the fetus proves to be affected by a serious inborn error of metabolism. (The method by which inborn errors of metabolism can be detected by prenatal diagnosis will be discussed further in Chapter 6.)

Summary

By the early 1950s, experiments with bacteria and viruses had made it clear that the genetic material is not protein, but DNA. The Watson-Crick model of DNA structure was proposed in 1953. The DNA molecule consists of four kinds of nucleotides joined to each other in a double-stranded helix in such a way that the sequence of nucleotides in one strand automatically determines the sequence of nucleotides in the other strand. DNA replicates by the separation of the two strands; each strand then serves as a template for the manufacture of a new, complementary strand.

Biochemically, genes are specific regions along DNA molecules. Most genes include coded information that specifies the exact order in which amino acid building blocks are linked together to form a specific protein. Genes that code for proteins are known as structural genes; they usually consist of coding regions (axons) and noncoding regions (introns). Proteins are made up of long chains of amino acids, and the function of a given protein depends on its precise three-dimensional shape. Three successive nucleotides in one strand of a DNA molecule code for one amino acid in a protein. Protein synthesis requires a second kind of nucleic acid, RNA, and the actual stringing together of amino acids into proteins takes place outside the nucleus on structures called ribosomes. Messenger RNA (mRNA) is single stranded and has a base sequence complimentary to one of the strands of DNA. Tranfer RNAs serve as adapters for translating codons in mRNA into amino acids in polypeptide chains.

Persons affected by sickle-cell diseases are homozygous for a gene that results in a single difference in the amino acid in the beta chain of their hemoglobin molecules compared with the normal beta chain. But the hemoglobin of persons who have sickle-cell trait (heterozygotes) is more resistant to malaria than is normal hemoglobin. This probably accounts for the widespread distribution of the allele for the abnormal beta chain in persons who live in areas where malaria is common, in spite of the reduced reproductive fitness of homozygotes. In one form of beta thalassemia, the defect in the genetic program occurs in one of the introns of the beta chain of hemoglobin.

Garrod coined the term *inborn errors of metabolism* for human abnormalities resulting from defects in genes that code for enzymes. Alcaptonuria, albinism, and PKU are caused by defects in enzymes that speed up rates of metabolism of the amino acids phenylalanine and tyrosine. The best way to deal with inborn errors of metabolism is to prevent them. This can be accomplished by identifying carriers by means of pedigree analysis or tolerance tests, and identifying affected fetuses at early stages of development by means of prenatal diagnosis.

Suggested Readings

Molecular Biology of the Gene, 3d ed., by J. D. Watson. W. A. Benjamin, 1976. An authoritative overview of the molecular genetics written to be understandable even to those who have little knowledge of chemistry.

"The Genetic Code: III," by F. H. C. Crick. *Scientific American*, Oct. 1966, Offprint 1052. A summary of how the sequence of base pairs in DNA provides information for the synthesis of specific protein molecules.

"The Visualization of Genes in Action," by O. L. Miller, Jr. *Scientific American,* Mar. 1973, Offpring 1267. With the aid of the electron microscope, one can see genes being transcribed into RNA and watch RNA being translated into protein.

"Split Genes," by Pierre Chambon. *Scientific American*, May 1981. Describes in some detail how specific nuclear enzymes edit and splice mRNA.

"Genetic Gibberish in the Code of Life," by Graham Chedd. *Science 81*, Nov. 1981. Popular-level article on the workings of intron and exons.

"The Processing of RNA," by James E. Darnell, Jr. *Scientific American*, Oct. 1983. Provides some of the details about mRNA after it is transcribed and before it leaves the nucleus.

"A New Understanding of Sickle Cell Emerges," by Thomas H. Maugh II. *Science*, vol. 211, 16 Jan. 1981. A review of recent research concerning what makes sickle cell adopt their abnormal shapes.

Recombinant DNA: A Short Course, by James D. Watson, John Tooze, and David T. Kurtz. Scientific American Books, 1983. An authoritative, up-to-date, and readable account of the basics of molecular biology and genetic engineering.

Chapter 4

The Regulation of Gene Expression

The body of an adult human being is made up of about 100 trillion cells—roughly 1000 times the number of stars in our galaxy. This enormous number of basic building blocks can be classified into about 100 kinds, each of which is specialized for performing a particular task of benefit to the entire organism. Thus, nerve cells (neurons) conduct electrical impulses, muscle cells shorten in length, sex cells transmit the genetic program to offspring, and so on. In general, what makes a particular kind of cell specialized is the distinctive set of proteins it synthesizes, in addition to the so-called housekeeping proteins that are found in nearly all cells. As you know, a fertilized egg contains the complete human genetic program, half of which comes from the mother and the other half from the father. During the enormously complex process of development, each newly formed cell also receives the complete genetic program, because before cell division takes place, the entire genetic program is replicated and then distributed equally between the two resulting cells. Yet very early in development, certain cells become specialized and begin to synthesize the distinctive sets of proteins that will characterize them in the adult. It has been estimated that a particular kind of cell expresses only a tiny portion—perhaps 100,000th—of the entire genetic program. What about the thousands of other genes that are present in each nucleus but not expressed? Are they destroyed or irreversibly switched off in most cell types, or is their expression somehow selectively inhibited in the huge numbers of cells in which their gene products are not usually produced?

In this chapter, our main concern is the regulation of gene expression—that is, the ways in which human genes are switched on and off. Because of recently developed techniques such as those employed in genetic engineering, knowledge of the control of gene expression is increasing enormously. Some of the proposed mechanisms to be discussed are already known to be important; others may turn out to be of little significance. Control mechanisms that influence transcription—the synthesis of an mRNA molecule on a DNA template—are the *primary* means of regulating the expression of human genes, and this is the major focus of this chapter. It should be noted, however, that gene expression can be regulated at any point along the complex path from mRNA synthesis to the stringing together of specific amino acids of a given protein.

We shall consider the roles of regulatory proteins and how they interact with specific base sequences in DNA, how the orientation of the backbones of the double helix may influence gene expression, and how one X chromosome in the body cells of females becomes inactivated. This chapter concludes with an examination of how human cancers can arise because of alterations in the human genetic program. Let us begin by considering how the expression of the gene for the beta chain of hemoglobin is regulated.

These flattened structures, which resemble stacks of elongated saltines, are cells from the lens of the human eye as seen through the scanning electron microscope (magnified 812 times). The distinctive shape of the cells depends largely on the distinctive proteins they contain. Within each cell, specialized protein molecules are also stacked up and aligned in layers, which allows the precise focusing of light rays on the retina. (From *Tissues and Organs: A Text-Atlas of Scanning Electron Microscopy*, by R. G. Kesseland and R. H. Kardon. W. H. Freeman and Company. Copyright © 1979.)

Fetal Hemoglobins and Regulatory Genes

As discussed in the preceding chapter, the protein portion of a hemoglobin molecule consists of two alpha chains and two beta chains. Each of these amino acid chains is coded for by a different region of nuclear DNA, that is, by a different structural gene. The gene that codes for the alpha chain is located on chromosome number 16; it becomes active very early in fetal development and remains so throughout adult life. But the gene that codes for the beta chain behaves differently. Although two beta chains are present in almost all of the hemoglobin molecules of normal adults, they are not present in the hemoglobin molecules of a developing fetus or of a newborn infant. The gene that codes for the beta chain becomes active during the second month of embryonic life, and some hemoglobin molecules made up of two alpha chains and two beta chains are present from then on. (As discussed in the following paragraphs, in normal adults such molecules of hemoglobin A comprise about 97 percent of the total hemoglobin.) Like the molecules of normal adults, the hemoglobin molecules of developing fetuses and of infants up to six months old contain two alpha chains, but most do not contain two beta chains. Instead of beta chains, most of the hemoglobin molecules of fetuses and newborn infants contain a protein known as the *gamma chain*, of which there are two slightly different kinds. Hemoglobin made up of two alpha and two gamma chains is known as *fetal hemoglobin*.

As shown in Figure 4-1, the number of hemoglobin molecules containing two gamma, rather than two beta, chains changes dramatically shortly after birth. The number of molecules containing gamma chains falls off in proportion to the increase in the number of molecules containing beta chains. By the age of one year, gamma chains are normally found in only about 1

4–1   The kinds of polypeptide chains in hemoglobin change during development. All hemoglobin molecules have 2 alpha chains. The curves indicate the percentage of the total number of hemoglobin molecules that contain each type of non-alpha chain at various stages of development. (After Huehns et al., "Human Embryonic Hemoglobins," *Cold Spring Harbor Symposia on Quantitative Biology*, 29, 1964.)

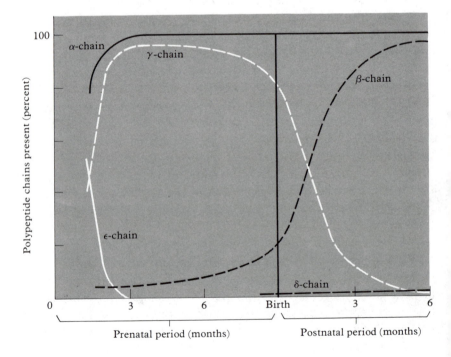

TABLE 4–1   NORMAL HUMAN HEMOGLOBINS

Chain	Combines with Two Alpha Chains and Four Heme Groups to Form:	When Present	Percentage of Total Adult Hemoglobin
Epsilon	Embryonic hemoglobin	Second and third months of fetal life	0
Gamma (two types)	Fetal hemoglobin (two types)	High fetal levels; drops to adult levels shortly after birth	1
Delta	Hemoglobin A_2	From about the third fetal month throughout adulthood	2
Beta	Hemoglobin A	From about the third fetal month; undergoes marked increase after birth	97

percent of hemoglobin molecules, and this percentage persists throughout adult life. Thus, the gene coding for the gamma chain is very active before birth, but becomes almost inactive by the end of the first year of life.

Very early in development, some of the hemoglobin molecules of fetuses contain still another chain known as the *epsilon chain*. The gene coding for the epsilon chain is active only during early embryonic life. As shown in Figure 4-1, hemoglobin molecules containing two alpha chains and two epsilon chains (known as *embryonic hemoglobin*) are present only during the first three months of development. Thus, by the end of the third month, the gene coding for the epsilon chain is shut off and never becomes active again.

Finally, shortly before birth the *delta chain* makes its first appearance. Hemoglobin molecules containing two alpha chains and two delta chains are called *hemoglobin A_2*, and they are never very numerous. In normal adults hemoglobin A_2 accounts for only about 2 percent of the total hemoglobin. The components of normal human hemoglobins are summarized in Table 4-1.

The genes that code for all of these amino acid chains are examples of structural genes because they bring about their effects by coding for a specific protein molecule. But not all genes are structural genes. From the study of the genetics of bacteria, we know that some genes, rather than coding for large protein molecules, have a controlling or regulatory role in protein synthesis. That this is probably also true of some human genes is suggested by certain genetic abnormalities of the kinds of hemoglobin discussed in Chapter 3.

In particular, some adults have been found to have abnormally high concentrations of fetal hemoglobin (two gamma chains). It has been suggested that a *regulatory gene*, may cause the rapid changeover from the synthesis of gamma chains to the synthesis of beta chains. Thus, those persons who have abnormally high concentrations of fetal hemoglobin can be though of as having a defective regulatory gene that failed to slow the production of gamma chains drastically.

Adults who have *only* fetal hemoglobin inside their red cells have been reported; surprisingly, this condition results in no apparent ill effects. This trait is autosomal recessive, and persons who are homozygous for it not only have no hemoglobin A (two beta chains), but also lack the small amounts of hemoglobin A_2 (two delta chains) that are normally present. It has been suggested that persons whose red cells contain only fetal hemoglobin have a defect in the regulatory gene that normally controls the function of the two structural genes for the beta and gamma chains. One reason for favoring this explanation is the recent discovery that the genes that code for the beta and beta-like chains of hemoglobin are arranged along the chromsome in the same order as they are expressed during development.

The cluster of structural genes responsible for the synthesis of the beta and beta-like chains is located on chromosome 11, and as just mentioned, the genes occur in the same order as they are expressed during development, namely, epsilon, gamma (two types), delta, and beta. The cluster consists of no fewer than 60,000 base pairs, yet the epsilon, two gamma, delta, and beta chains together are coded for by a total of about 3000 base pairs. This means that about 95 percent of the DNA of the beta-globin gene cluster does not code for protein. What does it do? Some of the noncoding DNA is accounted for by introns in the structural genes; some belongs to the category of DNA known as "repeated sequences", which will be discussed in the following chapter; and some of the noncoding DNA, especially that located between the structural genes, probably has a regulatory role and includes base sequences that promote or inhibit the expression of the structural genes within the cluster.

Turning on the Structural Gene for Fetal Hemoglobin in Adults

In December 1982, it was reported that certain adults with severe, incapaciting beta thalassemia or sickle-cell disease could be relieved of some of their symptoms by taking a potent anticancer drug that causes the normally inactive structural gene for the gamma chain of hemoglobin to be switched on. (How this is done is not yet known.) This treatment was attempted because it was known that those rare individuals who have only fetal hemoglobin as adults do not have anemia and lead normal lives. It was therefore reasoned that turning on the synthesis of the normally inactive gamma chain could help to relieve severe anemia, because the gamma chains would compete with the abnormal beta chains for the normal alpha chains, thus reducing the proportion of abnormal hemoglobin molecules. The drug in question, 5-azacytidine, was known to be capable of activating certain normally inactive genes in cells maintained in tissue culture; could it benefit these severely ill patients? The results were very impressive. The red blood cell count of the first patient treated increased more than 25 percent following one week of administration of the drug, and all patients treated showed a marked decrease in the number of abnormal red blood cells.

As shown in Figure 4-2, the structural formula of 5-azacytidine is very similar to that of cytidine, one of the four nucleotide building blocks of DNA. 5-Azacytidine is incorporated into newly synthesized DNA as readily as cytidine is, but DNA containing the drug has different chemical properties. The cytosine ring in cytidine is capable of being *methylated*; that is, a *methyl group*, with the simple chemical formula CH_3, can be added to one or more of its carbon atoms. This usually happens to the carbon atom labeled "5" in Figure 4-2; the result is 5-methylcytosine. This is important because *the addition of methyl groups to cytosine may be a major way of switching genes off.* DNA in the region of the human gamma globin gene contains large numbers of methyl groups in adults, but in fetuses it contains very few. It was suggested that 5-azacytidine brings about the activation of the gamma gene (among others) because it cannot be methylated; its carbon atom number 5 is already combined with a so-called aza group. In fact, it was found that the DNA in the region of the gamma globin gene in cells from the bone marrow of the treated patients is relatively poor in methyl groups (hypomethylated) following the administration of 5-azacytidine.

Recent research has disclosed that the gene for fetal hemoglobin is turned on by 5-azacytidine within one or two days, which is far too fast to be accounted for by hypomethylation of the DNA of the bone marrow cells that give rise to mature red blood cells. It has also been discovered that other anticancer drugs such as hydroxyurea, which do not result in the hypomethylation of DNA, can also rapidly turn on the gene for fetal hemoglobin. 5-Azacytidine is now thought to result in a rapid increase in the level of fetal hemoglobin not because it results in the hypomethylation of DNA, but because of its effects on the so-called stem cells in the bone marrow, which are the precursors of red blood cells. The gene for fetal hemoglobin is active in stem cells, but when these are transformed into mature red blood cells, the gene is somehow switched off (perhaps by the addition of methyl groups). 5-Azacytidine apparently causes stem cells to become "committed" to the production of hemoglobin before they are fully differentiated into red blood cells. Because the gene for fetal hemoglobin is active in stem cells, it is transcribed, and the level of fetal hemoglobin increases as more and more stem cells are affected by the drug.

Regulatory Proteins

Proteins that interact directly with portions of the double helix by binding to it, or which somehow influence its three-dimensional structure in a given region, are undoubtedly important in determining whether genes are expressed or not. When a structural gene is turned on, the double helix unwinds in a given portion, and one strand serves as a template for the synthesis of a complementary mRNA. The enzyme that transcribes mRNA from a DNA template is *RNA polymerase*. In order to initiate mRNA synthesis, RNA polymerase must become temporarily bound to a specific base sequence in DNA. Once it has "recognized" and become bound to this base sequence,

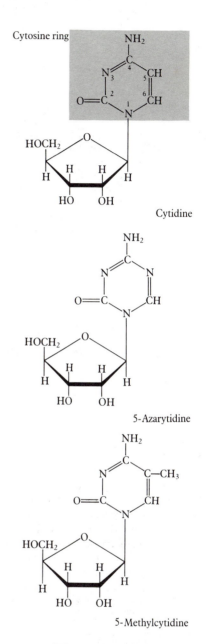

4–2   The structural formulas of cytidine, 5-azacytidine, and 5-methylcytidine. Within the cytosine ring (dark shading), the position of the atoms are numbered. Note that these three nucleotides differ in the atoms found at position 5 in the cytosine ring. (Structural formulas are discussed in Appendix II.)

the enzyme begins to travel along one strand of the double helix, synthesizing the mRNA as it goes. In recent years, it has become apparent that in order for human genes to be transcribed into mRNA, a region of DNA immediately "upstream" from the structural gene must first become relatively "relaxed" or uncoiled. It is at these relatively relaxed regions (often called "hot spots"), which are from about 100 to about 200 base pairs long, that RNA polymerase first becomes bound to the double helix. Within a given uncoiled region is a specific recognition site about 40 base pairs long, known as a *promoter*, which binds the RNA polymerase. Because the enzyme itself is a protein, it is not difficult to imagine how other proteins might interact with the uncoiled regions or promoters and block transcription. Alternatively, certain proteins could bind to DNA and cause certain regions to uncoil, thus exposing the promoters and favoring transcription. Both kinds of interactions of DNA and regulatory proteins are presumed to be important in determining which human genes are expressed and when, but direct evidence for human regulatory proteins is so far lacking. Most of what is known about regulatory proteins has been learned from the study of gene expression in bacteria.

Figure 4-3 illustrates an intriguing and economical control mechanism that so far has been demonstrated only in the genetic programs of bacteria. In

4–3   Regulation of the transcription of the structural genes (a, b, and c) that code for enzymes that metabolize lactose. Top, a structural gene, R, codes for a regulatory protein that binds to the operator sequence in DNA. Bottom, in the presence of lactose, the regulatory protein is inactivated because it becomes bound to lactose. This exposes the promoter sequence, and RNA polymerase moves in and begins to transcribe the mRNAs for enzymes a, b, and c.

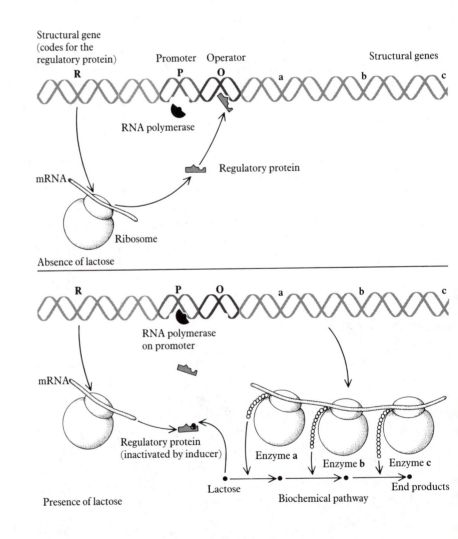

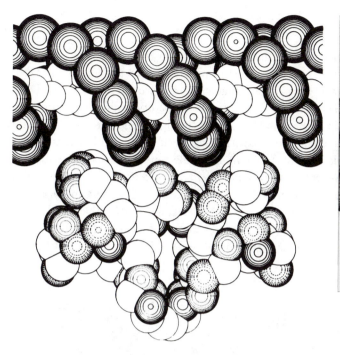

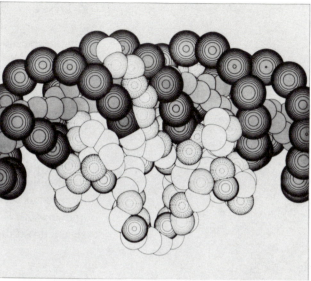

4–4 A computer-generated model of the three-dimensional structure of the cro regulatory protein. Grooves on the surface of the protein fit the spiral backbones of the double helix like pieces of a jigsaw puzzle. (Reprinted by permission from *Nature*, vol. 298, pp. 718–20. Copyright © 1982 Macmillan Journals Limited.)

this mechanism, the expression of two or more adjacent structural genes is controlled by another structural gene that codes for a regulatory protein, and by two nontranscribed regions of DNA: a promoter and an *operator*, which is about the same size as a promoter. The adjacent structural genes in this arrangement usually code for enzymes that are involved in the metabolism of a given substance, such as the sugar lactose. When no lactose is present, the regulatory protein binds to the operator sequence in DNA and thereby blocks the promoter site. Under these circumstances, RNA polymerase cannot bind to the promoter, so no transcription takes place. But when the cell is provided with abundant lactose, the genes that code for the enzymes are transcribed and the enzymes soon begin to accumulate inside the cell. What happens is this: Lactose combines with the regulatory protein that binds to the operator; when this happens, the regulatory protein can no longer combine with the operator, leaving the promoter site exposed. RNA polymerase then recognizes the promoter site, binds to it, and initiates the transcription of the structural genes that code for the enzymes that metabolize lactose. This efficient arrangement of structural genes and control elements appears to be unique to bacterial cells. In spite of extensive searching, to date no such arrangement has been documented in the human genetic program.

It is not difficult to imagine how a regulatory protein might interact with a DNA molecule. For many years, it was assumed that the protein probably had a shape that allowed it and a specific portion of DNA to fit together like pieces of a jigsaw puzzle. This assumption has now been confirmed, at least for regulatory proteins that inhibit protein synthesis. As shown in Figure 4-4, the shape of a protein known as cro, which turns off portions of the DNA of a certain virus allows it to nestle into the spiral grooves of the double helix, thus blocking transcription.

Z-DNA

Another recent discovery is that the shape of a regulatory protein known as *CAP* (catabolic gene activator protein), which turns on portions of the DNA of *Escherichia coli*, also has a shape that allows the protein to fit into the DNA like a key in a lock. But in this case, the shape of the protein is not complementary to that of the spiral grooves in the right-handed double helix, the form of DNA to which our discussion has been limited so far. It turns out that the two backbones of the DNA molecule can have several possible orientations. In the right-handed double helix, known as *B-DNA* (the form of the double helix first described by Watson and Crick), the two backbones form a smooth, continuous curve. In the left-handed form of the double helix, known as *Z-DNA*, the two backbones do not form a smooth curve but instead zigzag down the molecule, as shown in Figure 4-5. It turns out that *E. coli*'s

4–5   Right, the right-handed version of the DNA double helix, or B-DNA, is the form of the molecule first described by Watson and Crick. Left, in the left-handed form, known as Z-DNA, the two backbones zig-zag and do not form a smooth curve. (From Kolata, G., *Science*, vol. 214, 9 January 1981, p. 1108. Diagram by Alexander Rich. Copyright 1981 by The American Association for the Advancement of Science.)

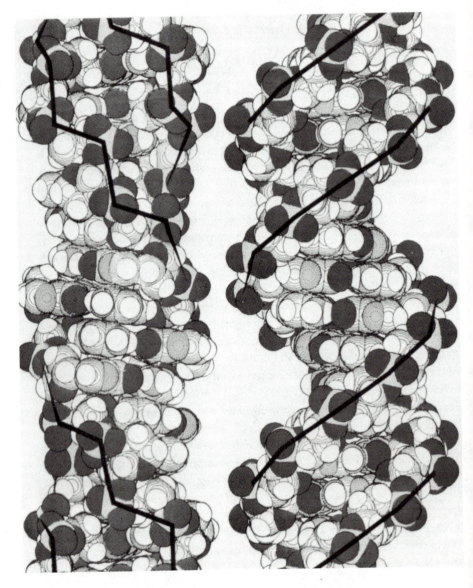

CAP protein, which stimulates transcription, has a shape that is complementary to a portion of Z-DNA. It is not known if the binding of the CAP protein causes a local portion of the double helix to assume the Z-DNA form or if CAP binds to DNA that is already in that form.

Z-DNA is fairly widespread in the human genetic program, and many biologists believe that it plays an important role in the regulation of gene expression. In addition to its possible regulatory effects by interacting with proteins, Z-DNA may bring about the transcription of certain genes because of the removal of methyl groups. In this hypothesis, naturally occurring Z-DNA is assumed to be genetically inactive and to be maintained in its zigzag shape by the methylation of cytosine in the zigzagged portions. When Z-DNA is demethylated, it assumes the B-DNA form. It is hypothesized that the switch from Z-DNA to B-DNA causes a distortion in the B-DNA at some distance from the recently demethylated stretch of Z-DNA. This distortion supposedly results in the unwinding of a specific area in B-DNA, thus exposing a promoter region that can be recognized by RNA polymerase. At any rate, although Z-DNA probably does play some role in the regulation of gene expression, its relative importance and mechanism of action remain to be worked out.

Inactivation of the X Chromosome

As you may recall from the discussion of Lyons' hypothesis in Chapter 2, very early in development one of the two X chromosomes in each body cell of a developing female becomes genetically inert. The inactive, tightly coiled X chromosome forms a dark-staining blob near the periphery of the nucleus—the so-called Barr body (see Figure 2-11).

What accounts for the (almost complete) inactivation of the X chromosome of normal females? Supercoiling of the DNA and proteins in the chromosome to form a very tightly packed lump undoubtedly plays a major role. Some experimental evidence also suggests that the addition of methyl groups to cytosine in the DNA of the X chromosome may be of some importance. In one experiment, several hybrid mouse-human cells were formed that contained a complete set of female mouse chromosomes and a single, inactive human X chromosome. The mouse cells were known to be deficient in an enzyme abbreviated HPRT, which is coded for by a gene on the X chromosome. When the hybrid cells were treated with 5-azacytidine, the drug that can be used to switch on the gene for fetal hemoglobin in adults, it was possible to detect some cells that contained HPRT; in addition, the enzyme was shown to be of the human type, which differs slightly from that of the mouse. Furthermore, several other human enzymes coded for by X-linked genes were also recovered from the hybrid cells. Because 5-azacytidine results in fewer methyl groups in DNA, these findings suggest that the addition of large numbers of methyl groups to cytosine within the double helix may be a major mechanism of X chromosome inactivation. On the other hand, a different set of experiments has shown no activation of an inactive human X chromosome when 5-azacytidine is added to cultured connective tissue cells from a normal female mouse. The researchers suggest that the foreign environment of the human X chromosome in the mouse-human hybrid cells results in far fewer methyl groups in the DNA, and that the addition of 5-azacytidine brings about the expression of X-linked genes only in DNA already relatively poor in methyl groups. The relative importance of methylation in the inactivation of the X chromosome is thus uncertain, and further evidence concerning its role is needed before the issue can be settled.

Oncogenes and Human Cancers

In recent years, it has become increasingly clear that many forms of human cancer can result from various kinds of genetic abnormalities, especially defects in regulating the expression of certain genes that control the rates at which cells divide. *Cancer* is an abnormal physiologic state in which affected cells undergo uncontrolled division and no longer confine themselves to orderly arrays in a given tissue or organ; instead, they form irregular masses of jumbled cells (*tumors*) and often invade and proliferate in tissues far from the origin of the first cancer cell. As shown in Figure 4-6, cancer cells in tissue culture have abnormal shapes, and instead of forming flat layers and maintaining a certain distance from neighboring cells, they pile up and roll over one another to form a chaotic array resembling the cells in a tumor. In addition to their uncontrolled division and irregular shapes, most cancer cells have distinctive metabolic abnormalities, altered cell membranes, and many other pathologic features. Yet in spite of their myriad abnormalities and great variety (over 100 kinds of human cancers are known), an increasing evidence suggests that many, perhaps most, human cancers arise from defects in the expression of a small number of genes that somehow control the rate of cell division.

In 1982, it was discovered that the cells of a certain kind of human bladder cancer contain a specific DNA sequence that, when isolated and introduced into the culture medium of cells in tissue culture, causes normal cells to be transformed into cancer cells. This specific stretch of cancer-causing DNA was treated with chemicals that unwind the double helix and allow the complementary strands to separate; it was then labeled by incorporating radioactive atoms into one of the strands. When the single-stranded, labeled DNA was added to a single-stranded copy of the entire human genetic program, the labeled DNA was found to bind very tightly to a specific stretch of normal human DNA. This implied that the cancer-causing DNA had a base sequence that was complementary, or nearly so, to a specific base sequence in the normal human genetic program. This conclusion was confirmed when the base sequence of the cancer-causing stretch of DNA, known as an *oncogene*, was compared to that of the corresponding normal gene, known as a *proto-oncogene*. The rather surprising result was that the oncogene and the proto-oncogene differ by a single base—a guanine in the normal gene is replaced by a thymine in the cancer-causing one. This particular proto-oncogene codes for a protein known as *p21*. The single base change in the oncogene results in a single amino acid substitution in the corresponding abnormal protein—a valine replaces the glycine usually found in position 12 of the polypeptide chain. Yet, in a way reminiscent of the disastrous effects of a single amino acid substitution in sickle-cell hemoglobin as compared to normal hemoglobin, the abnormal p21 somehow causes normal cells to be transformed into cancer cells. How this takes place remains to be worked out, but it is presumed that the normal p21 controls normal growth and cell division, whereas the abnormal one results in the loss of the normal control mechanism, as well as in the other typical features of cancer cells. To date, three other proto-

4–6   Top, when maintained in tissue culture, normal mouse connective tissue cells (fibroblasts) spread out to form a single layer in which each flattened cell keeps a certain distance from its neighbors. Bottom, when DNA extracted from a human bladder cancer is added to the cultured cells, the mouse cells are transformed into cancer cells. Note how the cancer cells round up and climb over one another; the rate of cell division is also markedly increased. (Courtesy of Erika A. Hartweig and Jonathan A. King, Biology Electron Microscope Facility, MIT.)

oncogenes and their corresponding oncogenes have been discovered in the human genetic program, and undoubtedly others await discovery.

In the case of the bladder cancer just discussed, the oncogene can be imagined as coming from a normal gene because of a *point mutation*, a single base change. Alterations in the base sequences of various proto-oncogenes in the human genetic program may explain the well-documented ability of certain chemicals, known as *carcinogens*, to cause cancer. Most carcinogens damage DNA molecules and ultimately result in DNA containing altered base sequences. Although the oncogene for human bladder cancer can transform normal cells in tissue culture into cancer cell, it is not known if the mere presence of the oncogene is sufficient to cause cancer in a human being. In fact, most cancer researchers maintain that the development of human cancers requires at least two different steps and perhaps more. It has recently been suggested that the development of cancer may require the activation of two different oncogenes, which correlates with the observation that the incidence of cancer is much higher in older people, whose DNA has been exposed to mutation-inducing environmental factors for many years.

The synthesis of an abnormal regulatory protein is not the only way in which oncogenes bring about their effects. A second way is by amplification of a proto-oncogene. The proto-oncogene is somehow replicated until it is present in multiple copies. The presence of these multiple copies causes the normal gene product of the proto-oncogene, a regulatory protein, to accumulate inside the cell, thereby transforming it into a cancer cell. A third way of activating proto-oncogenes is by changing their relative positions on a given chromosome, an important process that will be considered in the following chapter. Finally, cancers can also result from the introduction of proto-oncogenes or oncogenes into cells when they become infected by certain viruses.

Incorporation of Cancer-Causing Genes into the Genetic Program by Retroviruses

Let us conclude this chapter by considering how one important class of cancer-causing viruses, the *retroviruses*, bring about their effects. Like all viruses, retroviruses consist of a core of nucleic acid surrounded by a coat of protein. *In retroviruses, the nucleic acid is single-stranded RNA.* Once a retrovirus enters a cell, the protein coat is digested and the RNA is then used as a template for the synthesis of a complementary strand of DNA. The enzyme that brings about the transcription of the DNA from an RNA template is known as *reverse transcriptase*—"reverse" because the usual flow of genetic information is from DNA to RNA. The single-stranded DNA serves as a template for the synthesis of a complementary DNA strand, and the resulting doubled-stranded, viral DNA eventually becomes incorporated into the DNA of one of the host cell's chromosomes. A cell infected by a retrovirus thus ends up with an extra bit of DNA in its genetic program.

After the retroviral DNA has been incorporated into the chromosome, it is usually transcribed into RNA by enzymes of the host cell. The transcription of the viral DNA results in the synthesis of two different categories of RNA.

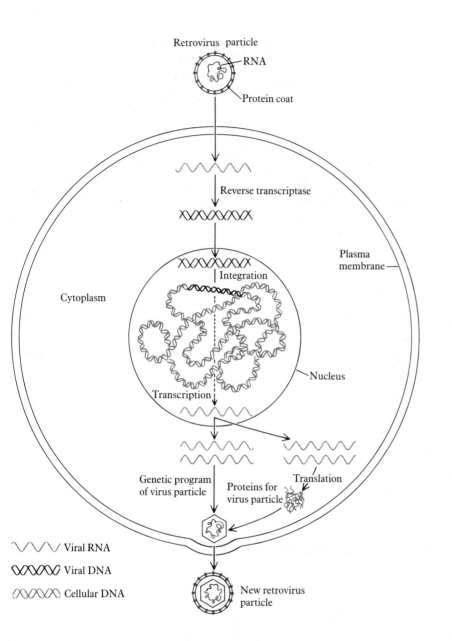

Retrovirus particle

RNA

Protein coat

Reverse transcriptase

Plasma membrane

Cytoplasm

Integration

Nucleus

Transcription

Genetic program of virus particle

Translation

Proteins for virus particle

∿∿∿ Viral RNA

⋊⋉⋊⋉ Viral DNA

⋀⋁⋀⋁⋀ Cellular DNA

New retrovirus particle

4–7   The infection cycle of a retrovirus begins when the viral particle is taken up by the host cell. The protein coat is digested, and virally derived reverse transcriptase then synthesizes viral DNA using the single-stranded RNA of the virus as a template. Viral DNA then enters the nucleus and becomes integrated into the DNA of the host cell. Transcription of the viral DNA produces RNAs for the synthesis of viral proteins as well as the complete stretch of single-stranded RNA that makes up the viral genetic program. The coat proteins automatically come together to form a protective shell surrounding the viral genetic program and a few molecules of other viral proteins, including reverse transcriptase. The newly formed viral particles are then released from the host cell to infect other cells.

First, several different mRNA molecules are synthesized and then translated by the host cell's ribosomes to produce viral proteins, including those of the coat and reverse transcriptase. Second, the entire genetic program of the retrovirus is transcribed into one long stretch of RNA that does not serve as an mRNA. Instead, the single-stranded stretch of RNA and a few protein molecules, including several of reverse transcriptase, become surrounded by a shell of viral coat proteins, resulting in a newly synthesized retrovirus. On their release from the host cell, the new retroviruses are ready to begin the infection cycle once again (Figure 4-7).

Retroviruses can cause cancer in two different ways. First, so-called weakly oncogenic retroviruses induce cancers in various animals after a long latent period lasting for many months. In this case, the DNA coded for by the viral RNA becomes inserted into the chromosome at a location that is very close to a naturally occurring proto-oncogene. It is thought that the transcription of the virally derived DNA (which ultimately results in the formation of new virus particles) causes the nearby proto-oncogene to be transcribed at a higher rate than in the absence of the viral DNA. The buildup of the regulatory protein coded for by the proto-oncogene presumably results in the transformation of the infected cell (Figure 4-8).

On the other hand, the so-called highly oncogenic retroviruses cause animals to develop cancer a few weeks after their cells become infected. *The DNA added to the genetic program during infection by highly oncogenic retroviruses contains a proto-oncogene or oncogene.* The cancer-causing genetic material is added to the viral genetic material during viral replication. The mRNA from an adjacent proto-oncogene or oncogene is incorporated into the stretch of single-stranded RNA that makes up the genetic program of the newly formed viruses; this process is not yet understood. In effect, the retroviruses have picked up a copy of a cancer-causing gene during the process of replication. When such highly oncogenic retroviruses infect other cells, the host cells contain an extra copy of a proto-oncogene or oncogene, which is transcribed whenever the viruses replicate. The result is either the accumulation of a normal regulatory protein or the introduction of an abnormal regulatory protein. Either way, infection by a highly oncogenic retrovirus soon pushes the host cell over the edge toward cancer.

More than 20 proto-oncogenes or oncogenes are known to be introduced into the genetic programs of various animals by means of highly oncogenic retroviruses. The number of weakly oncogenic retroviruses is much larger, and many more will probably be discovered in years to come. In humans, a retrovirus is known to cause one form of cancer of the white blood cells

4–8   Schematic diagram showing how a weakly oncogenic retrovirus known as ALV (avian leukemia virus), which causes leukemia in birds, is thought to produce its effects. DNA of viral origin is inserted into the host cell's DNA at a point very close to a cellular oncogene known as *c-myc*, which is usually not expressed in adults. When the viral DNA is transcribed, the *c-myc* gene is also transcribed into mRNA and then translated into a regulatory protein whose presence induces cancer in the host cell. (After *Recombinant DNA: A Short Course*, by James D. Watson, John Tooze, and David T. Kurtz, copyright © 1983. Scientific American Books.

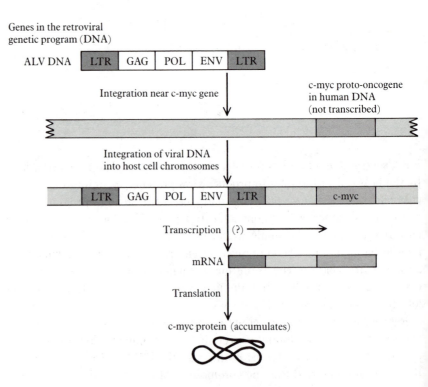

(leukemia), and several other kinds of blood cell cancers are suspected to be of retroviral origin. Although some retroviruses are downright treacherous because they can cause cancer, others are apparently harmless. Some retroviruses become incorporated into human chromosomes, direct the synthesis of new viral particles, and are released from the host cell without producing any obvious damage or disease. These curious entities probably evolved from mobile genetic elements in the normal human genetic program—the so-called jumping genes, which will be discussed in the next chapter. Very recently, apparently harmless retroviruses have been used to introduce new genes into the genetic programs of various kinds of cells, including human cells in tissue culture. As Chapter 6 will show, retroviruses may turn out to be important agents for the transfer of normal genes into the chromosomes of persons whose defective genes result in inborn errors of metabolism and other kinds of genetic diseases.

Summary

Each kind of cell in the human body expresses a distinctive set of structural genes and therefore has a distinctive set of proteins. Most of the genetic program is not expressed in a given cell type; many genes are expressed only during development and are never active in adults.

During fetal development, the composition of the hemoglobin molecule changes several times. Two alpha chains are always present, but the beta chain does not appear until the second fetal month. Before birth, epsilon, gamma, and delta chains are also present, and the order in which the beta-like chains appear is the same as the order of the structural genes on the chromosome. The changeover from one beta-like chain to another is probably controlled by a regulatory gene.

The drug 5-azacytidine can be used to switch on the normally inactive structural gene for the gamma chain of fetal hemoglobin in adults. The addition of methyl groups to cytosine in DNA, which in some kinds of cells is inhibited by 5-azacytidine, may be an important means of switching genes off.

Regulatory proteins are well known in bacteria and probably play an important role in the expression of human genes. RNA polymerase synthesizes mRNA on a DNA template; the enzyme recognizes and binds to specific promoter sequences in DNA. Many regulatory proteins have shapes that allow them to combine with DNA like pieces of a jigsaw puzzle.

The backbones of the double helix can exist in several forms. In B-DNA (the form of the molecule described by the Watson-Crick model), the backbones form a smooth curve, and in Z-DNA a zigzag pattern is present. Z-DNA, or the switch from the Z to the B form in a given region of the molecule, may be an important factor in the regulation of gene expression.

The mechanism whereby one of the two X chromosomes becomes inactive in the body cells of normal women is not known. The tight coiling of the DNA and proteins in the chromosome is of some importance, and the addition of methyl groups to DNA may also be a factor.

Oncogenes are genes that can cause cancer because of the presence of an abnormal regulatory protein, and proto-oncogenes can cause cancer because of the buildup of normal regulatory proteins that control the rate of cell division. Retroviruses, which may be either weakly or strongly oncogenic, employ the enzyme reverse transcriptase to synthesize DNA from an RNA template. Strongly oncogenic retroviruses add oncogenes or proto-oncogenes to the host cell chromosome, whereas weakly oncogenic retroviruses enhance the expression of nearby proto-oncogenes.

Suggested Readings

"The Visualization of Genes in Action," by O. L. Miller, Jr. *Scientific American*, Mar. 1973, Offprint 1267. With the aid of the electron microscope, one can see genes being transcribed into mRNA and watch mRNA being translated into protein.

"Chromosomal Proteins and Gene Regulation," by Gary S. Stein, Janet Swinehart Stein, and Lewis J. Kleinsmith. *Scientific American*, Feb. 1975, Offprint 1315. Discusses the structural and regulatory roles of the proteins in human chromosomes.

"Fetal Hemoglobin Genes Turned On in Adults," by Gina Kolata. *Science*, vol. 218, 24 Dec. 1982. Discusses in some detail the clinical trials in which 5-azacytidine was used in the treatment of severe beta thalassemia and sickle-cell disease.

"Hemoglobin—From F to A, and Back," by John W. Adamson. *The New England Journal of Medicine*, vol. 310, 5 Apr. 1984. Presents evidence that the hypomethylation of DNA is not responsible for the switching on of the gene for fetal hemoglobin by various anticancer drugs.

"Z-DNA: From the Crystal to the Fly," by Gina Kolata. *Science*, vol. 214, 4 Dec. 1981. A review of some of the structural features and possible biological functions of the left-handed form of the double helix.

"The DNA Helix and How It Is Read," by Richard E. Dickerson. *Scientific American*, Dec. 1983. Provides excellent illustrations of the structural differences between B-DNA and Z-DNA.

"The Molecular Basis of DNA-Protein Recognition Inferred from the Structure of CRO Repressor," by D. H. Ohlendorf et al. *Nature*, vol. 298, 19 Aug. 1982. Technical account of how the structure and mechanism of action of a regulatory protein were determined.

"Reactivation of an Inactive Human X Chromosome: Evidence for X Inactivation by DNA Methylation," by T. Mohandas et al. *Science*, vol. 211, 23 Jan. 1981. Details of the experiments with mouse-human hybrid cells discussed in this chapter.

"Studies of X Chromosome DNA Methylation in Normal Human Cells," by Stanley F. Wolf and Barbara R. Migeon. *Nature*, vol. 295, 25 Feb. 1982.

Provides experimental evidence that methylation is not an important factor in X chromosome inactivation.

"Variety in the Level of Gene Control in Eukaryotic Cells,"by James E. Darnell, Jr. *Nature*, vol. 297, 3 Jan. 1982. A technical article about the many mechanisms that control gene expression in nucleated cells.

"A Molecular Basis of Cancer," by Robert A. Weinberg. *Scientific American*, Nov. 1983. Discusses how human cancers can be initiated by oncogenes.

"Step by Step into Carcinogenesis," by John Cairns and Jonathan Logan. *Nature*, vol. 304, 18 Aug. 1983. A review of the evidence that the development of cancer occurs in at least two separate steps.

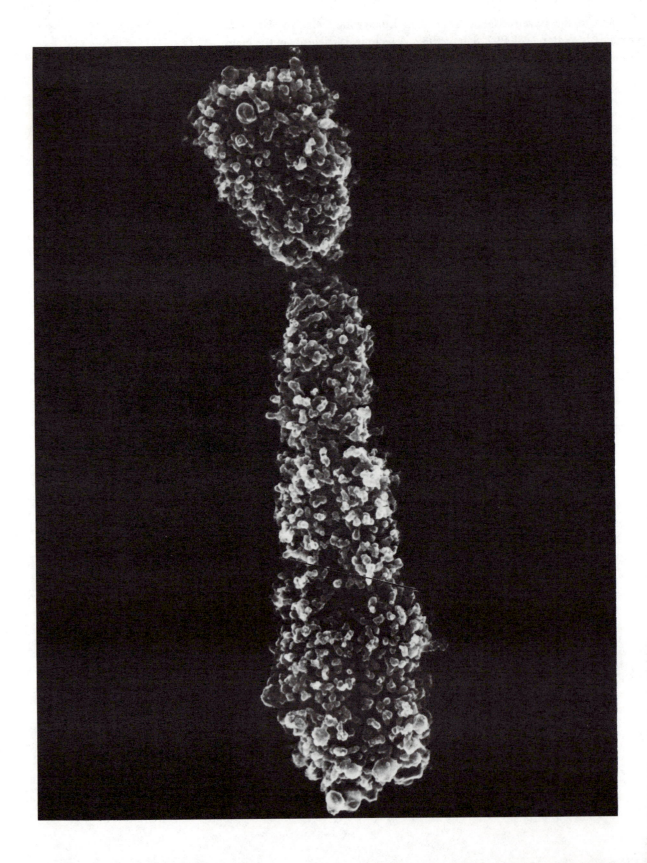

Chapter 5

The Human Genetic Program and Chromosome Map

The entire human genetic program consists of nearly 3 billion DNA base pairs and is divided into 46 segments, each of which is housed in a different chromosome. The DNA in each human chromosome is a single unique, double-stranded molecule that is closely associated with various kinds of proteins and is very tightly packed. In nondividing cells, chromosomes are still present in the nucleus, but they cannot be distinguished from one another because the DNA and its associated proteins exist in a much less tightly packed array than in the chromosomes of dividing cells. In the relaxed, unpacked state, the DNA molecule of an average-sized human chromosome is from 3 to 4 centimeters long. Placed end to end, the unpacked DNA in all 46 human chromosomes would stretch almost 2 meters (about 6.5 feet), yet each nucleus is so small that 200 of them could be lined up between two successive ridges on a human fingertip.

Working out the precise contents of such an extensive genetic program is indeed a formidable task, and only a tiny portion of the human genetic program has been deciphered so far. In recent years, this undertaking has become much easier because of powerful new techniques that allow the rapid and accurate sequencing of long stretches of DNA, the synthesis of highly precise molecular probes for locating specific base sequences, and the cloning of human genes in bacterial cells. However, although genes are now being precisely located within the human genetic program at the rate of several per month, the exact chromosomal location of less than 1 percent of the estimated total number of human genes is known.

In this chapter, our main concerns are the kinds of DNA found in the human genetic program, how DNA is packaged in chromosomes, and the various ways that human genes can be mapped—that is, assigned to a precise location on a particular chromosome. We shall also consider how the relocation of normal genes to abnormal sites because of chromosomal rearrangements is frequently associated with the development of cancer.

The Kinds of DNA in Human Chromosomes

Most of the DNA in the human genetic program is of unknown function. Nonetheless, several kinds (or categories) of DNA can be readily distinguished. We shall consider the main kinds of DNA in the human genetic program in order of decreasing abundance.

Long Unique Sequences

About 70 percent of human DNA is accounted for by relatively long stretches (from 5000 to 10,000 base pairs or longer) of unique base sequences that

Scanning electron micrograph of a hamster chromosome magnified 15,000 times. A chromosome from a dividing human cell would look very similar at this magnification. The knotted coils are highly condensed chromatin, which consists of equal amounts by weight of DNA and protein. (Courtesy Drs. Susanne M. Gollin and Wayne Wray, Kleberg Cytogenetics Laboratory, Department of Medicine and Department of Cell Biology, Baylor College of Medicine, Houston, Texas. © Susanne M. Gollin and Wayne Wray, 1983. All rights reserved.)

are present in a single copy. Included in this category are structural genes, which code for specific polypeptide chains, and regulatory genes, which bind enzymes or regulatory proteins and help to determine whether or not the structural genes are expressed. Based on the number of known metabolic pathways and the biochemical reactions that occur in human cells, and allowing for the regulatory genes that control the expression of structural ones, *it is estimated that the total number of functional genes in the human genetic program is between 50,000 and 100,000.* If we assume that the correct number is the larger one, and that an average gene (including the introns in the structural genes) is about 10,000 base pairs long, then we can account for about 1 billion DNA base pairs. But recall that there are nearly 3 billion DNA base pairs in the entire human genetic program. This means that about 2 billion base pairs, or roughly two-thirds of the DNA in the human genetic program, cannot be accounted for by structural genes or by known kinds of regulatory genes.

If only one-third of human DNA is accounted for by structural and regulator genes, what about the remaining 66 percent or so of the total human genetic program? Part of the answer can be obtained by studying Figure 5-1, a schematic drawing of the cluster of genes that code for the beta or beta-like chains of hemoglobin, which were discussed in the preceding chapter. As you can see, the structural genes that code for the polypeptide chains (including their introns) do not account for most of the base pairs in the gene cluster. Instead, most of the DNA is found in *spacer segments*, which do not code for protein and are not transcribed. Whether or not spacer DNA has some function remains to be seen. The beta-globin gene cluster also contains two so-called *pseudogenes*, which are the same base sequences as those for the beta chain, only without the introns. As shown in Figure 5-2, pseudogenes that are not transcribed into mRNAs may develop during infection with retroviruses, which introduce reverse transcriptase into human cells. The abundance of pseudogenes in the human genetic program is still unknown.

Moderate Repetitive Sequences

About 20 percent of the DNA in human chromosomes consists of moderate repetitive sequences, segments of DNA that are from about 130 to 300 base pairs long and can be present in thousands of copies. Some moderately repeated sequences, including those that code for the three kinds of RNA molecules found in ribosomes and those that code for the chromosomal proteins known as histones (to be discussed later), are arranged in tandem. The se-

5–1   Schematic drawing of the cluster of genes that code for the beta or beta-like chains of hemoglobin. Note that the structural genes are separated by spacer segments, which are not transcribed into mRNA. G_γ and A_γ represent two slightly different forms of the gamma chain, and $\psi\beta_1$ and $\psi\beta_2$ are pseudogenes, discussed in the text.

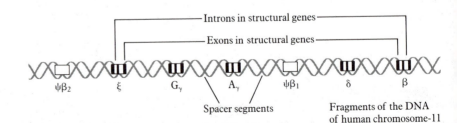

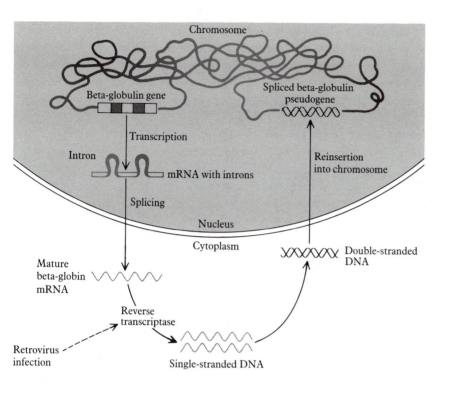

5–2 Pseudogenes have the same base sequence as a given structural gene, but they contain no introns. Pseudogenes may arise when a mature mRNA from the structural gene (an mRNA from which the transcribed introns have been spliced out) serves as a template for the synthesis of a complementary strand of DNA because of the action of reverse transcriptase introduced into the cell during infection by a retrovirus. After it is synthesized, the DNA produced by reverse transcriptase becomes inserted into the DNA of the chromosome, where it presumably becomes an untranscribed pseudogene.

quences coding for the ribosomal RNAs, for example, are present in about 200 copies, which are spread out in clusters among 10 different chromosomes. Each cluster consists of about 20 tandem copies of the genes that code for the ribosomal RNAs.

The most abundant of the moderate repetitive sequences in human DNA is known as the *Alu family*. These repeated sequences are about 300 base pairs long. As shown in Figure 5-3, all Alu sequences have certain base sequences in common, although there is some variation in the exact base sequences among members of the family. There are about 300,000 Alu sequences in human DNA; they are not found in tandem but are scattered throughout the other kinds of DNA—in spacer sequences, structural genes

5–3 The structure of members of the Alu family of repetitive sequences in human DNA. Two repeating sequences 40 base pairs long are present in all Alu sequences, each of which also has direct-repeat base sequences 7 to 10 base pairs long at each end. Elsewhere in the molecule, the base pair sequence can vary slightly in different members of the Alu family.

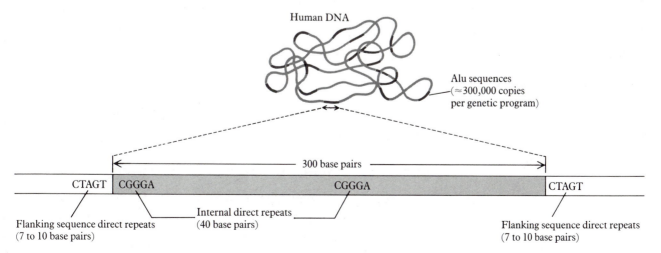

(including introns), and so on. Some Alu sequences are probably transcribed into small mRNAs that are abundant in human cells but are of unknown function. It has been suggested that Alu sequences may influence gene expression by switching nearby structural genes on or off, but their precise function or functions in the human genetic program are unknown.

Mobile Genetic Elements (Jumping Genes)

Mobile genetic elements, or so-called *jumping genes*, are surely present in the human genetic program. Although their numbers are probably not large, the consequences of their movement in the genetic material can be pronounced. The main reason for this is that their presence can markedly influence the expression of genes near the points where they enter or leave a given chromosome. If, for example, a jumping gene inserts intself into human DNA at a point near a proto-oncogene, the cancer-causing gene may be transcribed at a higher rate and cause the cell to be transformed into a cancer cell. In this regard, certain kinds of mobile genetic elements are very similar to retroviruses (discussed in Chapter 4). In fact, retroviruses are probably the descendants of a class of mobile genetic elements known as *transposons*. Retroviruses can be thought of as mRNAs of certain transposons that have become encapsulated by proteins and have left the genetic program to exist as infectious particles.

Several different kinds of mobile genetic elements have already been discovered. Their number and location can vary widely within the genetic programs of members of the same species. Some jumping genes consist of 10,000 base pairs; others are a few hundred nucleotides long. It is thought that all jumping genes have short, inverted repetitions of about 30 base pairs known as *insertion sequences* at each end. As shown in Figure 5-4, when a jumping gene enters the genetic program, it becomes integrated just after a specific "target sequence," which somehow becomes duplicated during the insertion process.

5–4   When a mobile genetic element becomes inserted into chromosomal DNA, it does so at a specific location—wherever a specific target sequence is found. The base sequence of the jumping gene is flanked by two oppositely oriented insertion sequences.

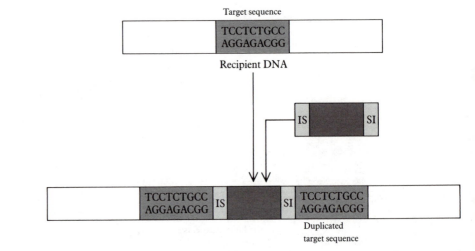

Mobile genetic elements were first described in the early 1950s by the American geneticist Barbara McClintock, whose experimental subject was the maize plant, or Indian corn. Although her careful crosses clearly indicated the existence of what she called "controlling elements," which could change their chromosomal locations, the idea that genes could move to various locations was at that time so unorthodox that it was not taken seriously. Like Gregor Mendel and his peas, Barbara McClintock and her Indian corn were ahead of their time. Since the late 1960s, a series of discoveries has convincingly demonstrated that mobile genetic elements are probably widespread in the genetic programs of most bacteria, plants, and animals, including humans.

Short Repetitive Sequences

About 6 to 10 percent of the human genetic program is accounted for by short repetitive base sequences, also known as *satellite DNA*. Short repetitive sequences are typically about seven base pairs long and are present in hundreds of thousands or millions of copies. Multiple copies of tandemly arrayed, short repetitive sequences are found at a specific location on each human chromosome. As shown in Figure 5-5, radioactively labeled mRNAs that correspond to the short repetitive sequences, which usually are not transcribed, bind to chromosomes in the region of the primary constriction, or centromere. The association of the short repetitive sequences with centromeres suggests that the sequences somehow play a role in the proper alignment and sorting out of chromosomes during cell division.

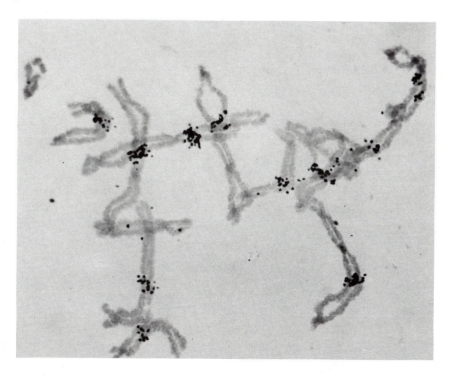

5–5 These salamander chromosomes have been incubated with synthetic, radioactively labeled mRNA molecules that have a base sequence complementary to that of short repetitive sequences in the salamander's DNA. The location of the radioactive mRNA was visualized by exposing the chromosomes to a photographic plate. Dark spots indicate the presence of labeled mRNA. As you can see, most of the complementary mRNA is located near the primary constriction or centromere of each chromosome—the point where the four arms of an X-shaped chromosome cross. (Courtesy of H. C. MacGregor.)

The Fine Structure of Human Chromosomes

In terms of end-to-end length, the unpacked DNA in a human chromosome is about 100,000 times longer than the chromosome in a dividing cell. By weight, human chromosomes are composed of roughly equal amounts of DNA and protein. It is primarily the association of DNA with the class of chromosomal proteins known as *histones* that allows the incredibly tight packing of the DNA in a fully condensed chromosome.

In nondividing cells, the nucleus contains an apparently structureless, coarsely granular material known as *chromatin*. When chromatin is studied under the electron microscope, it is seen to have a definite structure—it resembles beads on a string. Each "bead" of chromatin is known as a *nucleosome*. A nucleosome consists of a core, which is made up of two molecules each of four different kinds of histones, and the DNA is wound around the outside of the core. As shown in Figure 5-6, each nucleosome has two loops of DNA wound around it, and the nucleosomes are connected by short stretches of "linker DNA," which is the "string" connecting the beads. A fifth kind of histone is located on the outer surface of the nucleosome; it presumably helps to stabilize the nucleosome and may help to anchor the DNA in place.

The portion of DNA wrapped around a nucleosome is about 200 base pairs long and is about one-sixth of its unwound length. When cells prepare for cell division, the chromosomes become progressively more condensed. Exactly how the DNA and proteins in the chromosomes of dividing cells become so tightly packed is not known, but coiling and supercoiling are probably important, as illustrated in Figure 5-7.

The degree to which chromatin is condensed is an important factor in determining whether or not structural genes are expressed and regulatory

5–6   The core of a nucleosome consists of eight protein molecules—two each of four different kinds of histones (designated 2A, 2B, 3, and 4). A fifth kind of histone (1) binds to the outside of the core and may help to stabilize the DNA wrapped around the core histones. Adjacent nucleosomes are connected by linker DNA. (After *DNA Replication*; by Arthur Kornberg. W. H. Freeman and Company. Copyright © 1980.)

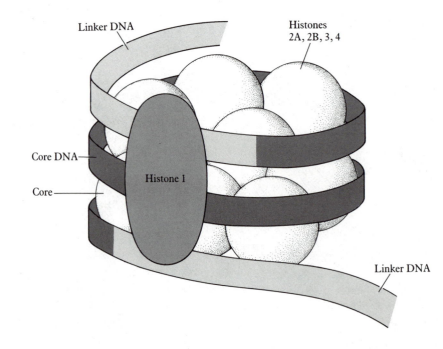

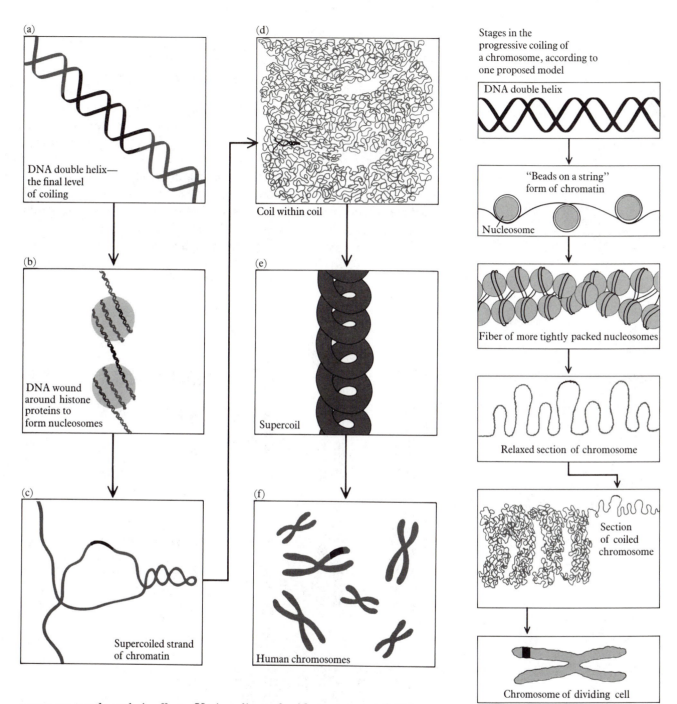

(a) DNA double helix—the final level of coiling

(b) DNA wound around histone proteins to form nucleosomes

(c) Supercoiled strand of chromatin

(d) Coil within coil

(e) Supercoil

(f) Human chromosomes

Stages in the progressive coiling of a chromosome, according to one proposed model

DNA double helix

"Beads on a string" form of chromatin

Nucleosome

Fiber of more tightly packed nucleosomes

Relaxed section of chromosome

Section of coiled chromosome

Chromosome of dividing cell

genes can produce their effects. Various lines of evidence suggest that genes located in relatively uncoiled, relaxed chromatin are more likely to be expressed than genes found in chromatin in a highly condensed state. The tightly coiled, highly condensed X chromosome that forms a Barr body in the nuclei of human females, for example, is genetically inactive, whereas its relaxed, more stretched-out counterpart contains hundreds of expressed genes. As the chromosomes become progressively more condensed prior to cell division, some regions in each chromosome become more tightly coiled than others. This can be demonstrated by staining the chromosomes at various stages of condensation. Selective staining reveals a specific *banding pat-*

5–7   The exact ways in which the "beads on a string" of relaxed chromatin become very tightly packed in the chromosomes of dividing cells remain to be determined. Coils and supercoils of the basic chain of beads on a string may be formed as shown here.

tern in which darker, more tightly coiled portions of chromatin alternate with lighter, less tightly coiled regions. As shown in Figure 5-8, the banding pattern of each human chromosome at a given stage of condensation is unique and allows every chromosome in the human genetic program to be positively identified. Although the chromatin in the lighter bands of the chromosomes of dividing cells is less tightly coiled, and therefore more likely to contain expressed genes than the darker, more highly coiled bands, no gene expression takes place during the complex process of cell division. Transcription into mRNA occurs only in nondividing cells with relatively relaxed chromatin. Nonetheless, there is convincing evidence that even in relatively relaxed chromatin, genes in the regions that become less tightly coiled in fully condensed chromosomes are more likely to be expressed than are genes in the regions that correspond to the most highly condensed portions of the chromosomes of dividing cells.

5–8   The banding pattern of specially stained human chromosomes from dividing cells reveals a unique pattern for each chromosome. Black bands represent areas of relatively tighter packing of DNA and proteins. (From "Standardization in Human Cytogenetics," by The Paris Conference, 1971. *Cytogenetics*, vol. 11, 313–362, 1972.)

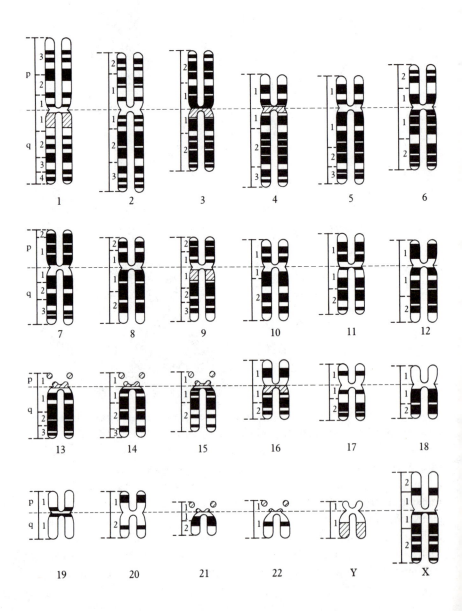

5–9   A close look at the dark, lacy structure at the bottom of this electron micrograph reveals that it is X-shaped. The X-shaped structure is a scaffold on which DNA and certain chromosomal proteins are tightly packed. Note the myriad loops of naked DNA that are attached to the scaffold. This chromosome has been depleted of the chromosomal proteins known as histones, which interact with the DNA to form a highly packed array in the chromosomes of dividing cells. (Courtesy of U. K. Laemmli, from Paulson, J. R., and Laemmli, U. K., *Cell*, vol. 12:817, 1977. © MIT.)

Histones are one of three major kinds of chromosomal proteins. The other two are *scaffold proteins* and *regulatory proteins*. As shown in Figure 5-9, histone-depleted human chromosomes have a rigid, chromosome-shaped scaffold which is made up of scaffold proteins, to which huge numbers of loops of naked DNA are attached. Regulatory proteins are a poorly understood group that probably coat the entire double helix, or at least those regions of DNA that contain structural and regulatory genes. It is presumably through the effects of chromosomal regulatory proteins that various genes are switched on or off during development.

Mapping Human Chromosomes

As was mentioned earlier in this chapter, the entire human genetic program is estimated to contain from 50,000 to 100,000 genes. Because there are only 23 pairs of human chromosomes, each chromosome (except for the Y chromosome) must contain a large number of genes. Parents contribute entire chromosomes to their offspring, so we would expect all of the genes on a given chromosome to be inherited en masse; such collections or sets of genes are known as chromosomal *linkage groups*. Thus, in humans, we would expect to find 23 groups of linked genes corresponding to the 23 pairs of chromosomes. This will surely turn out to be true, but our present knowledge of human linkage groups is meager. At present, nearly 4000 human genes are known, and about 600 of them can be assigned to specific chromosomes. Rarer still are those genes that have been precisely mapped, that is, pinpointed to an exact location on a specific chromosome. The precise chromosomal location of several hundred humans genes is now known, which amounts to less than 1 percent of the entire human genetic program. Nonetheless, progress in deciphering the human chromosome map has recently been greatly accelerated by the introduction of several new techniques. In this section, we shall discuss three methods that can be used to map human chromosomes—pedigree analysis, somatic cell hybridization, and *in situ* hybridization.

Pedigree Analysis

One of the main reasons that human linkage groups are so difficult to characterize is that most of the time the linkage of genes on a chromosome is not complete. What this means is well illustrated by considering data obtained from a cross of tomato plants that are homozygous recessive for two mutant alleles—*d*, which results in dwarf plants, and *f*, which results in fuzzy fruit. The homozygous recessive plants (*ddff*) were crossed with individuals that were heterozygous for the mutants alleles and were therefore of genotype *DdFf*, or "tall and smooth" as compared to the "dwarf and fuzzy" mutants. One such cross produced the following offspring:

161 *DdFf* (tall, smooth)
118 *ddff* (dwarf, fuzzy)
5 *Ddff* (tall, fuzzy)
5 *ddFf* (dwarf, smooth)

As you can see, the great majority of the offspring (279 out of 289, or 96.5 percent) have the same combination of traits as the parents. This can readily be explained by assuming that the traits are inherited together because they are located on the same chromosome. But what about the *recombinations* of traits—the approximately 3.5 percent of the offspring that have different combinations of the two traits from those of either parent? Linked traits recombine because *members of each pair of human chromosomes sometimes physically exchange sections with one another, thus unlinking traits that are found together on the same chromosome and allowing the generation of new combinations.* This exchange usually occurs during the formation of sex cells by a type of cell division known as *meiosis*, and the exchange of sections between chromosomes of a given pair is known as *crossing over*.

In meiosis, cells divide in such a way as to reduce the number of chromosomes in the nucleus by half. Meiosis is thus often referred to as *reduction division*, and in humans it occurs only during the production of sex cells. On the other hand, when a somatic (body) cell divides, the two cells that are produced have the same number of chromosomes as the original cell. This is because somatic cells duplicate their entire set of chromosomes before they divide and then distribute a complete set to each of the two cells produced by division. This type of cell division is called *mitosis*. But in meiosis, one cell divides twice and produces four cells, each of which has only half of a complete set of chromosomes, one member of each pair.

The most important features of meiosis in relation to crossing over and recombination can be summed up as follows: Like a somatic cell, a cell undergoing meiosis duplicates its entire set of chromosomes before it divides the first time. But before the duplicated sets of chromosomes separate in the first division of meiosis, they do something generally unheard of in chromosomes of somatic cells. Before the first division, most chromosomes undergoing meiosis become closely aligned with one another in matching pairs along their entire length. As shown in Figure 5-10, it is during this time of close alignment that the physical exchange of sections of matching pairs of chromosomes becomes apparent. As they separate during the first division, paired meiotic chromosomes are connected to one another at X-shaped areas called *chiasmata* (singular, *chiasm*), which presumably are points at which crossing over has occurred. Then, after the duplicate pairs have fully separated and formed two cells, each newly formed cell divides again to produce a total of four cells, each of which has half of a complete set of chromosomes.

The link between recombination and chromosome mapping is this: Genes farther away from one another cross over more frequently than those closer together, and *the frequency of recombination among the offspring is proportional to the physical distance between the two genes on the chromosome*. Thus, if crossing over between two genes occurs about 1 percent of the time, then the two are separated by one "map unit" of distance; in the example of the tomato plants mentioned above, the mutant alleles are separated by three and a half map units, and so on. The farther away the two genes are, the greater the number of crossovers. In fact, if two genes are separated from one another by more than 50 map units, recombinations may occur so frequently as to suggest that the genes are not linked, but located on different chromosomes.

Linkage and recombination are not easy to demonstrate in human pedigrees. This is because the data concerning two traits are usually not clearcut and common enough to allow us to decide whether the combinations observed in a given lineage are best explained by assuming that the traits are linked and that they undergo recombination by crossing over a certain percentage of the time. This is especially true of traits determined by linked genes located on an autosome. Even if pedigree analysis clearly indicates that two autosomal traits are linked, this indicates nothing about the particular autosome on which the genes that determine the traits are located. The particular autosome can be determined if the frequencies of the linked genes correspond to the presence of an abnormal chromosome, but they very rarely do. On the other hand, it is relatively easy to determine if two linked genes are located on the X chromosome, because such genes have the distinctive

5–10   The process of meiosis results in the formation of four cells, each of which has half as many chromosomes as the parent cell. The parent cell accomplishes this by duplicating its set of chromosomes once and then dividing twice. The chromosomes are duplicated just before the onset of the stage of mitosis known as prophase, and they first become visible as threadlike filaments. Each duplicated chromosome is actually double-stranded, and each strand is known as a sister chromatid. During prophase, matching pairs of double-stranded chromosomes become closely aligned with one another and segments of the double-stranded matching pairs may be physically exchanged by crossing over. During anaphase I the pairs become separated. One double-stranded member of each pair is distributed to each of the two cells that result. Each of these cells thus contains half as many double-stranded chromosomes as the parent cell. Then, without duplicating its genetic material, each of these cells divides again (anaphase II). During the second division, the two sister chromatids that make up each chromosome separate and each strand itself becomes a single-chromatid chromosome in one of the four resulting cells.

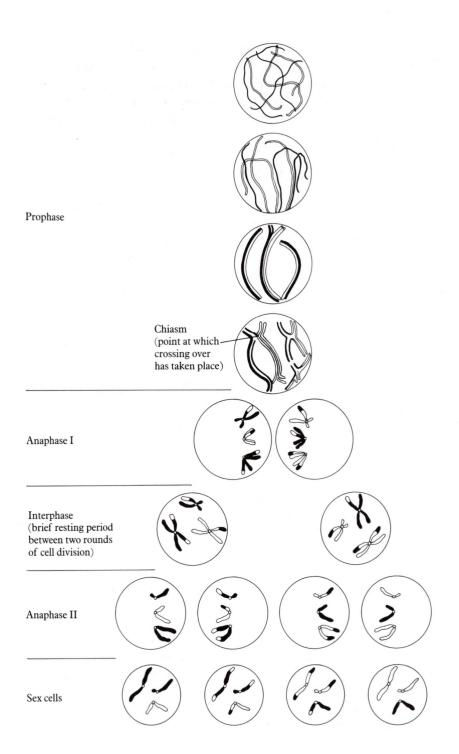

Prophase

Chiasm
(point at which crossing over has taken place)

Anaphase I

Interphase
(brief resting period between two rounds of cell division)

Anaphase II

Sex cells

pattern of inheritance discussed in Chapter 2. This accounts for the disproportionate number of mapped human genes that are X-linked. Although the X chromosome is thought to contain about as many genes as an autosome of similar size, about 160 of the nearly 500 human genes whose chromosomal locations are now known have been mapped on the X chromosome (Figure 5-11).

Somatic Cell Hybridization

In the early 1970s, a technique known as *somatic cell hybridization* was first used to figure out which genes are located on which chromosomes. This technique consists of fusing body cells (not sex cells) from other mammals, especially mice, with human body cells. As shown in Figure 5-12, human connective tissue cells (fibroblasts) are mixed with mouse tumor cells and are then incubated with various chemicals and a certain virus than enhances the fusion of the two kinds of cell to form "hybrid" cells. When the hybrid cells divide (by mitosis), the human chromosomes are progressively lost, apparently at random, and clones of cells that contain a full complement of mouse chromosomes and a few or a single human chromosome are produced. Under such circumstances, it is sometimes possible to detect the presence of enzymes or other proteins known to be produced by humans but not by mice. One can then conclude that the structural gene that codes for the protein is located on the particular human chromosome present in the hybrid cell. The chromosome can be determined by staining it and observing its banding pattern. The technique of somatic cell hybridization has so far resulted in the assignment of more than 100 human genes to specific chromosomes.

5–11   About one-third of the nearly 500 human genes mapped so far are located on the X chromosome. The reason for the apparently large number of X-linked genes is that they are relatively easy to identify and localize.

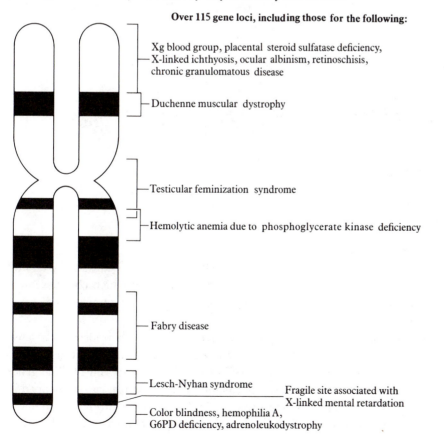

Over 115 gene loci, including those for the following:

Xg blood group, placental steroid sulfatase deficiency, X-linked ichthyosis, ocular albinism, retinoschisis, chronic granulomatous disease

Duchenne muscular dystrophy

Testicular feminization syndrome

Hemolytic anemia due to phosphoglycerate kinase deficiency

Fabry disease

Lesch-Nyhan syndrome

Fragile site associated with X-linked mental retardation

Color blindness, hemophilia A, G6PD deficiency, adrenoleukodystrophy

5–12   The technique of somatic cell hybridization. Human fibroblasts are mixed with mouse tumor cells and incubated with chemicals and Sendai virus particles, which enhance the fusion of the two kinds of cells. Hybrid cells result and they can be separated into clones, each of which has special properties, depending on which chromosomes it contains. As the cells undergo successive rounds of division, the human chromosomes are progressively lost until only one or a few remain in each clone of cells. (From "Hybrid Cells and Human Genes," by Frank H. Ruddle and Raju S. Kucherlapati. Copyright © 1974 by Scientific American, Inc. All rights reserved.)

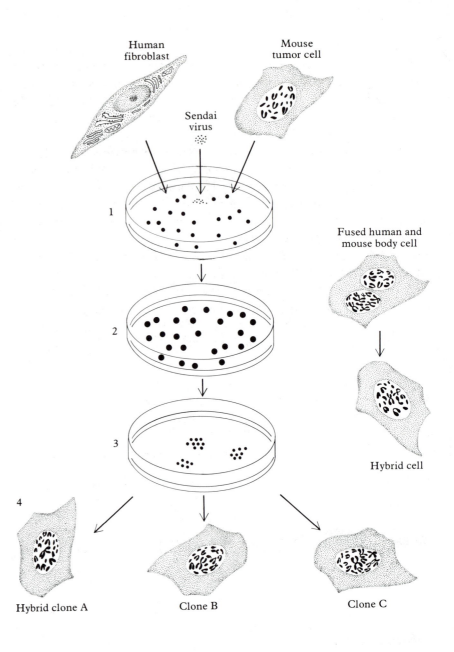

In Situ *Hybridization*

In the early 1980s, a powerful new technique was devised that for the first time allowed investigators not only to determine on which chromosome a given gene is located, but also to figure out exactly where on the chromosome it resides. The technique, known as *in situ hybridization* (*in situ* means "in place"), has even broader applications than somatic cell hybridization because it can pinpoint the position of a given gene whether or not that gene is expressed in the particular kind of cell involved in the experiment. The genes for the alpha and beta chains of hemoglobin, for example, can be mapped in chromosomes from connective tissue cells, which are relatively easy to maintain in tissue culture but which do not normally synthesize hemoglobin. The technique begins with the isolation of a specific mRNA coded for by a structural gene whose chromosomal location is to be mapped. Reverse transcriptase, the enzyme found in retroviruses, is then added to the specific mRNA and synthesizes a stretch of DNA that has a base sequence complementary to that of the mRNA. DNA produced in this way in known as *complementary DNA*, or *cDNA*. The specific cDNA is then produced in larger quantities by cloning in bacterial cells, a process to be discussed in the following chapter. Next, the cloned cDNA molecules are labeled by having radioactive atoms incorporated into them, and the labeled cDNA molecules are then added to the chromosomes of dividing cells, which have been stained in such a way that the banding pattern of each chromosome can be distinguished. Because the labeled cDNA molecules have a base sequence complementary to a portion of one of the strands of the double helix, the cDNA becomes tightly bound to that portion of the chromosome containing the base sequence of the gene in question. As shown in Figure 5-13, labeled cDNAs can thus

5–13   Schematic diagram showing the use of *in situ* hybridization to map the location of the gene for human alpha interferon, a protein that helps to protect cells against infection by viruses and perhaps against transformation into cancer cells. cDNA probes were prepared by exposing the mRNA for alpha interferon to reverse transcriptase; the probes were then mass-produced by cloning them in bacterial cells, and were then labeled by incorporating radioactive atoms into the molecules. When added to cells undergoing mitosis, the labeled probes became attached to the chromosomal location where the gene for alpha interferon is found. This location was revealed by exposing the labeled chromosomes to a photographic plate. Each dot reveals the presence of a labeled cDNA probe. Although there is some scatter among the dots (which may indicate that the gene for alpha interferon is present in several copies or may merely be an artifact introduced by chance), the results of many such labeling experiments clearly show that the gene for alpha interferon is located on the short arm of chromosome-9. (After Recombinant DNA: A Short Course, by James D. Watson, John Toose, and David T. Kurtz, copyright © 1983. Scientific American Books.)

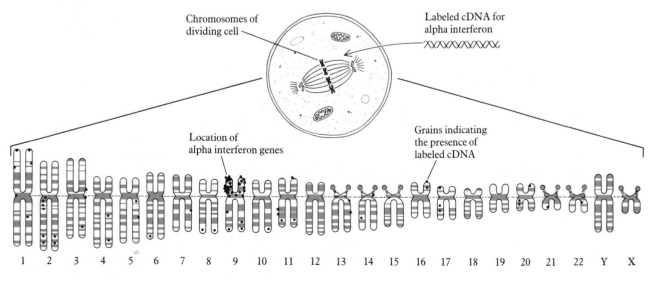

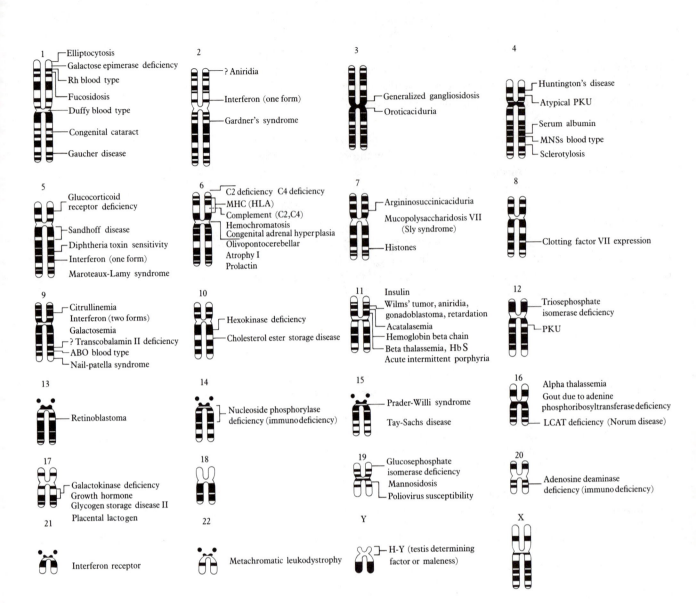

5–14   Representative genes of the human chromosome map. Most of the genes code for genetic diseases with simple Mendelian patterns of inheritance. Some of the mapped genes are normal genes that are of medical interest because they determine blood groups, maleness, and so on. After "The Anatomy of the Human Genome," by V. A. McKusick. *Hospital Practice*, vol. 16, 4, 1981.

serve as precise molecular probes for localizing specific base sequences in the human genetic program.

The techniques of pedigree analysis, somatic cell hybridization, *in situ* hybridization, and others (including restriction fragment length polymorphisms, discussed in the following chapter), have together allowed the assignment of more than 600 human genes to specific chromosomes and the precise mapping of several hundred mutant and normal genes. Figure 5-14 shows the chromosomal locations of some representative genes of the human genetic program.

Why go to the trouble of mapping human chromosomes? First, detailed maps will become more and more essential for the accurate diagnosis of many kinds of genetic diseases, especially in fetuses known to be at high risk for a given disorder. Second, as will be discussed in the following chapter, detailed chromosome maps will one day allow researchers to pinpoint the exact locations for introducing normal genes into human chromosomes that contain defective genes. A third reason is that chromosome mapping provides valuable information concerning the regulation of gene expression, especially in various kinds of cancer. And fourth, as a leading investigator has put it, the mapping of human chromosomes is like climbing Mount Everest. It is exciting because of the enormity of the challenge.

Chromosomal Rearrangements and Human Cancers

As techniques for mapping human chromosomes and for visualizing their unique banding patterns have become progressively more precise, a new understanding of human cancer that was only vaguely hinted at a decade ago has emerged. It is this: *The abnormal cells in most forms of human cancer have a chromosomal defect.* Although a wide range of chromosomal defects may be involved, two kinds predominate in human cancer cells. The first is the deletion of a specific band from a given chromosome; the second is reciprocal translocation—the exchange of segments between specific chromosomes that break at precise, characteristic points.

The deletion of a specific band or segment of a given chromosome is observed primarily in cancer cells from solid tumors, such as those arising in the bladder, kidney, ovary, breast, lung, large intestine, and retina. A good example is the tumor known as *retinoblastoma*, a form of eye cancer that originates in cells of the retina, usually before an affected child is 5 years old. (Provided the condition is detected early enough, 85 percent of children with retinoblastoma can be cured, often without the loss of the affected eye.) Retinoblastoma is usually transmitted as an autosomal dominant trait, and it frequently arises sporadically by mutation in previously unaffected family lines. In many cases, retinoblastoma is associated with the deletion of the midportion of a certain band on chromosome 13. The deletion is usually found only in the tumor cells, not in normal body cells. A similar situation is observed in many cases of Wilms' tumor, a cancer of the kidney, which also appears during infancy or early childhood. In this case, the deletion is in chromosome 11. Although the exact mechanism by which the deletion of a specific segment of DNA results in cancer is yet to be determined, it is assumed that the deleted DNA and its related proteins somehow prevent the expression of a nearby oncogene.

Whereas a specific deletion of the genetic program is usually associated with solid tumors, reciprocal translocation is usually associated with cancers of blood cells, of bone marrow, or of organs of the lymphatic system, especially the lymph nodes and spleen. For example, in at least 95 percent of adults with a certain form of leukemia (chronic myelogenous leukemia), reciprocal translocation occurs between chromosomes 9 and 22. One form of lymph node cancer known as *Burkitt's lymphoma* deserves special mention. Most cases of Burkitt's lymphoma are associated with a reciprocal translocation

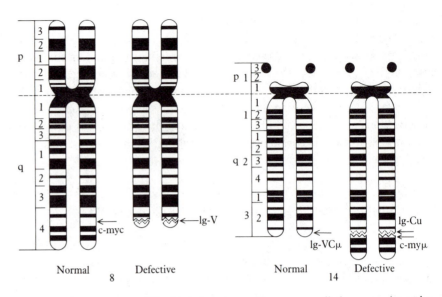

Normal   8   Defective   Normal   14   Defective

5–15   In most cases of Burkitt's lymphoma, the cancer cells have a reciprocal translocation between the ends of the long arms of chromosomes 8 and 14. The translocation results in the placement of a human proto-oncogene (*c-myc*) very close to a gene (Ig-Cμ) that codes for part of an antibody molecule and is translated at a high, fairly constant rate. These chromosomes were stained at the so-called 1200 band stage, when 1200 separate light and dark bands can be distinguished in human chromosomes. The numbers on the left indicate how the bands are labeled; *p* designates the short arm of a chromosome, *q* the long arm. (From "The Chromosomal Basis of Human Neoplasia," by Jorge J. Yunis. *Science*, vol. 221, p. 227, 15 July 1983. Copyright 1983 by The American Association for the Advancemebnt of Science.)

between chromosomes 8 and 14, and the breakpoints always occur at the same locations. This is intriguing because, as shown in Figure 5-15, the portion of chromosome 8 that is translocated to chromosome 14 is known to contain a proto-oncogene, which is usually not expressed in adults. Furthermore, the proto-oncogene-bearing portion of chromosome 8 becomes attached to chromosome 14 very close to a structural gene that codes for one of the polypeptide chains found in certain kinds of antibodies. This is important because the structural gene in question is usually transcribed at a

5–16   The arrows indicate the chromosomal locations of seven known proto-oncogenes in the human genetic program. Several proto-oncogenes have also been assigned to chromosome 20 and 21 but their precise locations have not yet been mapped. (After *Recombinant DNA: A Short Course*, by James D. Watson, John Tooze, and David T. Kurtz, copyright © 1983. Scientific American Books.)

Chromosome number

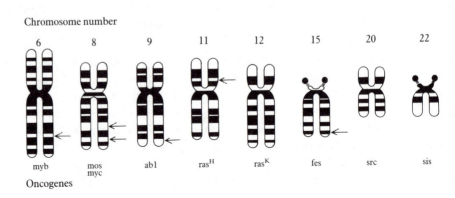

Oncogenes

high, relatively constant rate. The translocation of a proto-oncogene to a region of the genetic program where transcription is carried out more or less continuously may result in the transcription of the proto-oncogene as well. In this regard, the effects of translocation are similar to those of certain cancer-causing viruses, which were discussed in Chapter 4.

Deletions and reciprocal translocations are not the only chromosomal abnormalities observed in human cancer cells. Extra or missing entire chromosomes are sometimes involved, and inversions (regions in which the DNA of a given segment is the reverse of normal) have also been reported. As more data accumulate concerning the specific chromosomal defects associated with the approximately 100 different kinds of human cancers, chromosomal regions containing known proto-oncogenes will be carefully studied to see if they consistently change their position or are found close to altered or deleted chromosomal segments. As shown in Figure 5-16, several human proto-oncogenes have already been precisely mapped; others have been assigned to a given chromosome. The possibility that relocated, expressed proto-oncogenes are involved in a wide variety of human cancers is suggested by the observation that the same chromosomal defect is associated with two to five different kinds of cancer. Mapping of human chromosomes may eventually reveal that relatively small numbers of a few kinds of chromosomal abnormalities involving nearby proto-oncogenes are associated with most human cancers.

Summary

The human genetic program is composed of several kinds of DNA. Long unique sequences make up about 70 percent of the total and include structural and regulatory genes. Moderate repetitive sequences, such as the Alu family, are widepsread in human DNA but of unknown function. Mobile genetic elements are surely present in the human genetic program, but their function and significance remain to be determined. Short repetitive DNA sequences are clustered near the primary constriction, or centromere, of each chromosome and play a role in the proper sorting out of duplicated chromosomes during cell division.

The tight packing of DNA in the chromosomes of dividing cells depends mainly on proteins known as histones. In nodividing cells, the nucleus contains a granular material called chromatin, whose structure resembles beads on a string. Each bead, or nucleosome, is made up of a core of histones around which the DNA is wrapped several times. Genes located in relatively uncoiled, relaxed chromosomal regions are more likely to be expressed than are genes in tightly coiled, highly condensed regions. The banding pattern of each human chromosome is unique and allows the identification of even isolated chromosomes. Chromosomes also contain a rigid framework made up of scaffold proteins, and the outer surface of the chromosome is probably coated with various kinds of regulatory proteins.

Human chromosomes can be mapped by observing the frequency of recombination of linked traits in human pedigrees. Recombination occurs by crossing over, the exchange of corresponding sections between chromosomes of a given pair, usually during the production of sex cells by meiosis. Somatic cell hybridization can be used to assign genes to a particular chromosome, provided the gene is expressed in the kind of human cell that is fused with mouse cells to produce hybrids. *In situ* hybridization utilizes labeled cDNA as a probe to localize the exact position of a given gene on a given chromosome. At present, more than 600 human genes have been assigned to specific chromosomes, and several hundred have been mapped to particular chromosomal locations.

The abnormal cells in most forms of human cancer have a chromosomal defect. Deletions of portions of chromosomes may allow the expression of nearby proto-oncogenes and are consistently observed in the chromosomes of human cancer cells from solid tumors. Cancers of blood cells and bone marrow are characterized by cells with reciprocal translocations of specific segments of particular chromosomes. In Burkitt's lymphoma, one of the translocated chromosomal segments contains a proto-oncogene that may be activated when it is placed near a gene that is usually translated at a high and fairly constant rate.

Suggested Readings

"Repeated Segments of DNA," by R. J. Britten and D. E. Kohne. *Scientific American*, Apr. 1970. Compares the abundance of unique and repeated DNA sequences in the genetic programs of living organisms ranging from bacteria to humans.

"A Naturalist of the Genome," by Roger Lewin. *Science*, vol. 222, 28 Oct. 1983. Describes the research by Barbara McClintock that led to the discovery of mobile genetic elements (and to the Nobel Prize for Physiology or Medicine in 1983).

"Do Jumping Genes Make Evolutionary Leaps?" by Roger Lewin. *Science*, vol. 213, 7 Aug. 1981. A speculative article concerning the possible role of mobile genetic elements in the evolution of new species.

"The Nucleosome," by Roger D. Kornberg and Aaron Klug. *Scientific American*, Feb. 1981. Discusses the detailed structure of nucleosomes and how they are arranged in human chromosomes.

"The Anatomy of the Human Genome," by Victor A. McKusick. *Hospital Practice*, Apr. 1981. A very readable update on the status of the human chromosome by the acknowledged expert in the field.

"A New Therapeutic Territory: Gene Mapping," by Joseph Hixson. *Therapaeia of Medical Tribune*, Feb. 1981. Describes some of the newer techniques that can be used to map human chromosomes.

"Hybrid Cells and Human Genes," by Frank H. Ruddle and Raju S. Kucherlapati. *Scientific American*, July 1974. How the fusion of human somatic cells with the cells of other mammals can yield information that can be used in mapping human chromosomes.

"Replication Timing of Genes and Middle Repetitive Sequences," by Michael A. Goldman et al. *Science*, vol. 224, 18 May 1984. Provides evidence that certain middle repetitive sequences play a regulatory role during development.

"The Chromosomal Basis of Human Neoplasia," by Jorge J. Yunis. *Science*, vol. 221, 15 July 1983. A technical article that summarizes the specific chromosomal abnormalities associated with a wide range of human cancers.

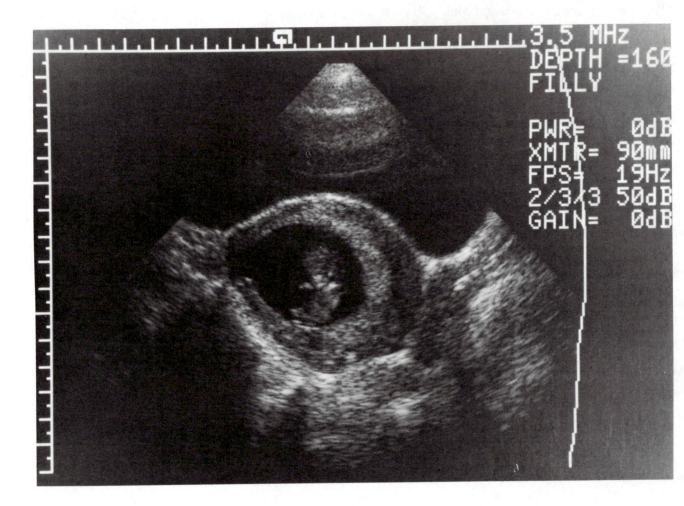

Chapter 6

Prenatal Diagnosis and Genetic Counseling

The birth of a severely deformed or genetically abnormal infant is one of life's greatest emotional traumas. As of the mid-1980s, about 4 percent of all babies born in the United States will have some kind of birth defect. This means that about 250,000 infants with significant birth defects are born each year.

Not surprisingly, once an affected infant has been born, the parents and other relatives usually seek professional counseling to ascertain the risk of recurrence of the abnormality in their future offspring. When the abnormal trait is determined by a single defective gene that has a simple Mendelian pattern of inheritance, the recurrence risk is relatively easy to estimate. Anyone who has read the first two chapters of this book should be able to determine the odds. But it is one thing to know the percentages and quite another to be willing to take the risk. Consider the case of familial retinoblastoma, a malignant eye tumor of children that is inherited as an autosomal dominant trait. Adults who had the eye tumor as children but were cured by surgery, radiation, and other forms of treatment have a 50-percent chance of passing on the abnormal gene to their sons and daughters. The decision on whether or not to have children is made more agonizing by the fact that there is a 50:50 chance that the child will be entirely normal (with regard to retinoblastoma) and will never have to know the pain and at least partial loss of vision, as well as the seemingly endless series of examinations and treatments that occupied so many of the childhood hours of the affected parent. On the other hand, this relatively optimistic viewpoint is offset by the fact that not all cases of familial retinoblastoma, even if detected early, can be cured. The decision on whether or not to have children would be somewhat easier to make if there were a way of telling if a fetus had inherited the normal or abnormal gene, and if the parents were willing to abort an affected fetus. (In the case of retinoblastoma, such a test is not yet available but probably will be soon.) Of course, the decision to proceed with the abortion of a fetus that has been shown to have inherited a debilitating abnormal gene (or an abnormal set of chromosomes) is by no means easy or without emotional impact. But for those individuals who accept the abortion of genetically abnormal fetuses, tests that are performed before birth—in other words, *pre-*

This ultrasound photograph shows a normal fetus during the first trimester of pregnancy, about 10 weeks after conception. The fetus is a mere 30 millimeters (0.12 inch) long. (Courtesy of Dr. Roy A. Filly, Chief, Section of Diagnostic Ultrasound, University of California, San Francisco.)

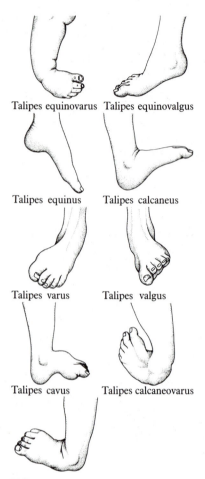

Talipes equinovarus   Talipes equinovalgus

Talipes equinus   Talipes calcaneus

Talipes varus   Talipes valgus

Talipes cavus   Talipes calcaneovarus

Talipes calcaneovalgus

6–1   Clubfoot is a congenital deformity of one or both feet in which the foot is twisted out of shape or position. Many varieties of clubfoot (technically known as *talipes*, from the Latin word for "clubfoot") are known, but the so-called equinovarus variety, in which the heel is turned inward and the foot is extended, accounts for 95 percent of all cases. The next most common is the calcaneovalgus variety, in which the heel is turned outward and the toes are elevated. The remaining kinds of clubfoot are rare.

natal diagnosis—can relieve the emotional and financial burdens of providing medical treatments for a severely ill or deformed child whose chances of survival to adulthood are often very poor. For those prospective parents for whom abortion is unacceptable, prenatal diagnosis can determine whether the fetus is normal or affected, and thus allow the parents to prepare themselves to take care of an affected child.

In this chapter, our main concern is with techniques of prenatal diagnosis and how the information obtained from them can be used in genetic counseling. Because of the recent introduction of new techniques such as gene cloning, and because of the development of new methods for obtaining fetal samples for analysis, the field of prenatal diagnosis is growing very rapidly, and new discoveries will probably continue to be made at an increasing rate for some time. nonetheless, as we shall see, our overall ignorance of the subject vastly exceeds our knowledge. Let us begin the discussion of prenatal diagnosis by considering a group of genetically determined abnormalities that do not usually follow simple Mendelian patterns—the so-called structural malformations.

Multifactorial Nature of Most Structural Malformations

About 150 kinds of structural malformations have been officially catalogued by the birth defects branch of the Centers for Disease Control. The most common of the major structural malformations include (1) cleft lip, which may be accompanied by cleft palate; (2) clubfoot, in which one or both feet are twisted out of shape or position (Figure 6-1); (3) dislocation of the hip in males; (4) pyloric stenosis, found mostly in males, in which the first part of the small intestine is markedly narrowed; and (5) various kinds of neural tube defects (discussed below). In the United States, structural malformations are by no means rare. They are observed in two or three out of every 100 births.

Like most other human traits, structural malformations do not have simple Mendelian patterns of heredity, but they do tend to run in certain families. As shown in Table 6-1, the brothers and sisters of an infant with one of the five birth defects mentioned in the preceding paragraph are more likely to be affected than the general population. The rapid fall-off in risk between brothers and sisters as compared to nieces and nephews can be explained by assuming that the structural malformation depends on *many* genes (and on certain environmental factors). Because distant relatives have only a small portion of genes in common with an affected individual, they are less likely to inherit all of the genes that together result in a given structural malformation. Evidence for the *polygenic* nature of the inheritance of cleft lip and palate comes from measurements of the widths of the midfaces of mothers, fathers, sisters, and brothers of affected individuals. All of these relatives tend to have wide midfacial areas. It is thought that some sort of dosage or threshold effect influencing several different genes is involved because an affected infant results only if enough of the genes in question are inherited. It is also assumed that environmental factors are important in determining if a particular structural malformation develops. No such factors have yet been iden-

TABLE 6-1   FAMILY PATTERNS IN SOME COMMON CONGENITAL MALFORMATIONS

	Cleft Lip or Cleft Palate or Both	Talipes Equinovarus	Hip Dislocation in Males	Pyloric Stenosis	Spina Bifida with Anencephaly
General population	1:1000	1:1000	2:1000	5:1000	8:1000
First-degree relatives	40:1000	25:1000	50:1000	50:1000	56:1000
Second-degree relatives	7:1000	5:1000	6:1000	25:1000	—
Third-degree relatives	3:1000	2:1000	4:1000	7.5:1000	—

SOURCE: From "Genetics of Common Disorders," by C. O. Carter. *British Medical Bulletin*, vol. 25, 1969, p. 54.

tified for most birth defects, but as will be discussed later, the lack of a certain vitamin is thought to be an important factor in the development of some kinds of neural tube defects.

About 70 percent of structural malformations are of unknown cause and are presumed to be *multifactorial* in that they depend on many genes and on various environmental factors. Thirty percent of these structural malformations can be traced to a specific cause: 20 percent are associated with single-gene defects; 5 percent are attributed to chromosomal abnormalities; from 2 to 3 percent to exposure to drugs or environmental chemicals; from 2 to 3 percent to maternal conditions, such as severe diabetes or infections of the uterus; and a final 1 percent to irradiation received during pregnancy.

Genetic counseling for individuals related to someone with a known structural malformation varies depending on whether or not the exact genetic basis is known. If the malformation is associated with a known single-gene defect or with exposure to a certain drug, the recurrence rates for various relatives can be predicted accurately. Most of the time the exact genetic basis is not known, and counselors must rely on statistics concerning the observed recurrence rates among relatives, such as those given in Table 6-1. Prenatal diagnosis is available for many single-gene defects and for all chromosomal abnormalities, which together account for about 25 percent of·all structural malformations. Another group of structural malformations—neural tube defects—can also be detected by prenatal diagnosis. Neural tube defects are so prevalent and serious that they demand further discussion.

Prenatal Diagnosis of Neural Tube Defects

Neural tube defects result from abnormal development of the embryo's neural tube, which gives rise to the brain and spinal cord. In the United States, neural tube defects occur in about 0.3 percent of all pregnancies and are of two main types, which may occur together. First, *anencephaly* results from the failure of the neural tube to give rise to the cerebral hemispheres and other higher centers of the brain. Anencephalic fetuses also usually lack the bones of the skullcap; their nearly empty skulls stop developing at the level

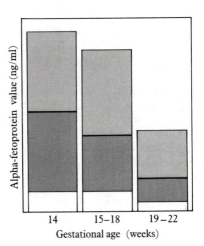

6–2   The level of alpha-fetoprotein in the fluid surrounding the fetus changes during the course of normal pregnancy. Elevated levels may be associated with the presence of a neural tube defect. (From "Prenatal Diagnosis of Genetic Defects," by Renata Laxova, *Postgraduate Medicine*, vol. 65, no. 3, March 1979.)

of their eyebrows and contain no higher brain centers. The second major neural tube defect, *spina bifida*, usually results from the abnormal development of the tail end of the neural tube. In this malformation, part of the lower end of the spinal column fails to develop, and nerve tissue of the spinal cord in that region is either exposed on the body surface or protrudes from the base of the spine.

A couple that has one infant with a neural tube defect has a 5 percent chance of having another; for couples with two affected children, the risk of recurrence is from 10 to 12 percent. In the early 1970s, when the technique of amniocentesis (discussed in the following section) had just come into widespread use, it was observed that the amniotic fluid surrounding fetuses with neural tube defects contained relatively large amounts of a protein known as *alpha-fetoprotein*. Alpha-fetoprotein is produced by the fetal liver; it is present in all developing fetuses and in the fluid that surrounds them in the womb, but in fetuses with neural tube defects the protein levels are much higher. Since the mid-1970s, the level of alpha-fetoprotein in the fluid surrounding embryos between 14 and 16 weeks of gestation has been used to diagnose neural tube defects in pregnancies known to be at increased risk. More recently, screening tests have been devised that measure the levels of alpha-fetoprotein in the blood of women in the second trimester of pregnancy (at about 18 weeks of gestation). If the level of alpha-fetoprotein in the pregnant woman's blood is higher than normal, amniocentesis is performed in order to measure the level in the fluid surrounding the fetus. As shown in Figure 6-2, the level of alpha-fetoprotein in the fluid surrounding normal fetuses follows a definite pattern. If abnormally high levels are observed, a neural tube defect should be strongly suspected. Unfortunately, a high level of alpha-fetoprotein in the maternal blood or in the fluid surrounding the fetus is not an infallible indicator of a neural tube defect. Elevated levels also occur in multiple pregnancies (twins, triplets, and so on) and in fetuses with abnormal liver function. Because amniocentesis is now always preceded by an ultrasound scan of the pregnant uterus (see below), the presence of two or more fetuses can be easily detected, and anencephaly and many cases of spina bifida can be confirmed. The widespread availability of second trimester screening of maternal blood for elevated levels of alpha-fetoprotein may soon help to reduce the incidence of neural tube defects.

In addition to the screening of maternal blood for the level of alpha-fetoprotein, two other factors may help to reduce the overall incidence of neural tube defects. First, beginning about 1930, there was an unexplained spontaneous decline (independent of prenatal diagnosis) in the incidence of neural tube defects. In fact, the incidence of anencephaly and spina bifida in the United States and the United Kingdom is now about half of what it was a decade ago. The reason for the decline is unknown (but welcome). Second, there is good preliminary evidence suggesting that neural tube defects may depend in part on a relative lack of the vitamin folic acid. This connection was first hinted at by the observation that pregnant cancer patients who received a drug that blocks the metabolic activities of folic acid frequently underwent spontaneous abortion of fetuses with neural tube defects. Studies employing laboratory animals have clearly indicated a relationship between the lack of folic acid and the presence of neural tube defects, and several studies in which pregnant women received supplements of folic acid have suggested a protective effect. Carefully designed investigations are now underway that should soon help to clarify the role of folic acid and the relationships between genetic and environmental factors in the development of neural tube defects.

The Use of Amniocentesis to Detect Several Kinds of Genetic Abnormalities

Amniocentesis is the insertion of a long, thin needle into the fluid-filled membranous sac, or *amnion*, surrounding the fetus, followed by the withdrawal of some of the fluid (Figure 6-3). The *amniotic fluid* in which the fetus is suspended is produced by the fetus itself and has several sources. Some of the fluid is secreted by the upper respiratory tract and some diffuses through the fetal skin or the umbilical cord, but the most important source of amniotic fluid is fetal urine. Amniotic fluid contains several kinds of cells of fetal origin. Included are cells that slough off from the fetus' skin, from the lining of the respiratory and urinary tracts, from the umbilical cord, and from the inner wall of the amnion (which, like the fetus itself, is derived from the fertilized egg, not from maternal tissues).

Amniocentesis is usually performed in the 14th to 16th week of pregnancy (16 to 18 weeks after the woman's last menstrual period), when about 200

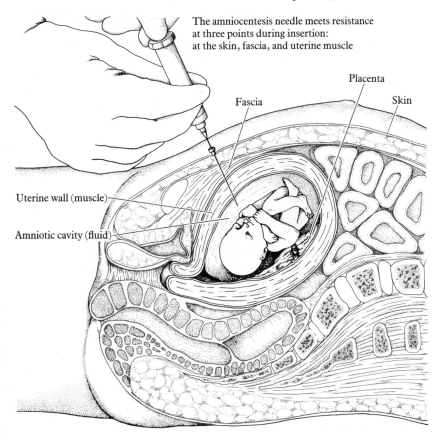

The amniocentesis needle meets resistance at three points during insertion: at the skin, fascia, and uterine muscle

Placenta

Fascia

Skin

Uterine wall (muscle)

Amniotic cavity (fluid)

6–3   Amniocentesis is the withdrawal of some of the amniotic fluid surrounding the fetus during the second trimester of pregnancy. Local anesthesia is used because the needle must pass through the mother's abdominal wall as well as through the wall of the uterus. The procedure is usually performed on an outpatient basis.

TABLE 6–2   INDICATIONS FOR AND INCIDENCE OF
ABNORMALITIES DETECTED BY AMNIOCENTESIS IN A
COMBINED SERIES OF 8181 SUBJECTS*

	Subjects		Abnormalities	
Indications	Number	Percentage	Number	Percentage
Cytogenetic study				
Maternal age ≥ 35 years	3224	39.4	89	2.8
Previous trisomy-21	1562	19.1	22	1.4
Parental translocation	211	2.6	16	7.6
Other chromosomal abnormality in family	967	11.8	16	1.7
X-linked disorder (sex determination)	334	4.1	158 (78)†	47.3 (23.3)
Metabolic disorders	331	4.0	76	23
Neural tube defects	1519	18.6	87	5.7
Others	33	0.4	0	
	8181	100.0	464	5.7

SOURCE: From "Prenatal Diagnosis of Genetic Defects," by Renata Laxova. *Postgraduate Medicine*, 3 Mar. 1979, p. 248.
* Canadian series of 1020, U.S. series of 1040, and European series of 6121 subjects.
† Figures in parentheses are the actual number and percentage of abnormalities detected, in contrast to the number of male fetuses.

milliliters (7 ounces) of amniotic fluid are present. The needle is inserted into the amniotic sac under local anesthesia, usually preceded by an ultrasound scan of the pregnant uterus. Sound waves of high frequency (from 20,000 to several million cycles per second) are aimed at the fetus, reflected, and electronically transformed into an image of the fetus on a video screen (frontispiece). The ultrasound scan locates the fetal head, the placenta (which connects the fetal circulation to that of the mother and bleeds profusely if pricked by a needle), and any pockets of easily accessible amniotic fluid. (As mentioned earlier, the ultrasound scan can detect many kinds of neural tube defects and multiple pregnancies.) When the sample of amniotic fluid has been obtained, a small portion is set aside for direct biochemical analysis, and the rest is put into tissue culture bottles and placed in an incubator to allow any cells of fetal origin to begin to multiply. In order to obtain enough dividing cells to permit chromosome analysis and metabolic studies, the cells must be kept in tissue culture for about three or four weeks. This means that the results of amniocentesis are usually not available until from 17 to 20 weeks into the pregnancy, at which time the woman is noticeably pregnant and can feel the movement of the fetus in her womb. The risk that the mother will

TABLE 6–3   FREQUENCIES AND RECURRENCE RISKS OF
SOME COMMON CHROMOSOMAL ANOMALIES

Syndrome	Chromosome Pattern	Frequency in Live-Born Infants*	Recurrence Risks to Sibs
Trisomy-21 (Down's)	47,+21	1:770	About 1%
Trisomy-18	47,+18	1:6600	About 2 to 3%
Trisomy-13	47,+13	1:5000	Negligible
Turner's	45,XO	1:5400	Negligible
XXX female	47,XXX	1:1000	Negligible
Klinefelter's	47,XXY	1:1100	Negligible
XYY male	47,XYY	1:900	—

* SOURCE: From *Birth Defects Atlas and Compendium, The National Foundation-March of Dimes*, by D. Bergsma (ed.). Williams & Wilkins, 1973.

develop major complications from amniocentesis is negligible, and the risk of inducing an inadvertent abortion is also very low—about 0.25 percent.

In the United States, approximately 30,000 women undergo amniocentesis each year. Although amniocentesis is safe and widely available, it is estimated that as of the mid-1980s, only about 10 percent of the women who are at substantial risk for having a genetically defective child have had the procedure. The percentage of high-risk women who are screened will probably increase as more and more disorders become detectable by amniocentesis. In large medical centers in the United States, the incidence of genetic abnormalities detected by amniocentesis has consistently been about 5 percent. This figure indicates that 95 percent of the women considered to be at risk are reassured that their offspring does not have the particular abnormality for which the amniocentesis was performed. This conclusion is, of course, a great relief for the prospective parents, but they must bear in mind that at present, ultrasound scans and amniocentesis can detect only a limited number of genetic abnormalities and that the overall risk that any American newborn will have a significant birth defect is about 4 percent.

Table 6-2 summarizes the reasons for which amniocentesis is performed in the United States, Canada, and Europe and the relative frequencies of the most important detectable disorders. We have already discussed the roles of ultrasound scanning and amniocentesis in the detection of neural tube defects. In the next section, we shall consider the role of amniocentesis in the prenatal diagnosis of two other kinds of genetic disorders: chromosomal abnormalities and inborn errors of metabolism.

Chromosomal Abnormalities
In the United States, chromosomal abnormalities are observed in one out of every 200 births. Table 6-3 lists the frequencies and recurrence risks of the most common disorders that result from abnormal numbers of chromosomes.

The most frequently encountered chromosomal abnormality is Down's syndrome (trisomy-21), which was discussed in Chapter 1. As you may recall, the incidence of Down's syndrome increases markedly with the age of the mother. Amniocentesis is therefore recommended for mothers 37 years old or more (some experts recommend that screening begin at age 35). The incidence of Down's syndrome among the offspring of older mothers is as follows: one in 100 for women in their middle thirties, one in 30 for women in their early forties, and one in 10 for women in their middle forties (see Figure 1-20). Amniocentesis is very accurate in the detection of abnormal numbers of chromosomes or parts of chromosomes. As discussed in the first two chapters, abnormal numbers of autosomes result in severe genetic defects; most affected individuals have short life spans and cannot take care of themselves. For those couples who accept abortion, amniocentesis is a reliable means of reducing the incidence of these devastating defects.

By contrast, for certain abnormalities in the number of sex chromosomes, the decision on whether to proceed with an abortion is not so easy. For example, women with only one X chromosome (Turner's syndrome) may be of normal intelligence and may have nearly normal life spans. In the case of male fetuses with the sex chromosome composition XYY, the situation is even more complex. XYY men are more likely than men with normal sex chromosomes (XY) to engage in behavior that results in their being apprehended and placed in a mental-penal institution. Nonetheless, in spite of the risk of criminal behavior, about 96 percent of XYY males do not come into conflict with the law. As will be discussed in Chapter 8, the behavioral consequences of sex chromosome constitution XYY are uncertain and very controversial. The decision on whether to have an abortion when amniocentesis reveals this chromosomal abnormality is not clear-cut.

Inborn Errors of Metabolism

In the United States, inborn errors of metabolism due to a single defective gene (for which the affected individual is usually homozygous) are observed in one or two out of every 100 births. Over 600 known genetic diseases that have a recessive pattern of inheritance result from single-gene defects. Of these 600 diseases, the exact defect—usually the presence of an abnormal enzyme or another kind of protein—is known for about 200. In sickle-cell disease, the defect is in the beta chain of hemoglobin; in PKU, a defective enzyme is present, and so on.

In principle, if the exact nature of the genetic defect is known, it should be possible to devise a test to detect heterozygotes. This is desirable because when two heterozygotes mate, there is a 25 percent chance that they will have an affected offspring, and most inborn errors of metabolism are devastating, untreatable diseases that result in death in infancy or childhood. If the defect can be detected in heterozygous carriers of the disease, then it should also be detectable in fetal cells obtained by amniocentesis, provided that the defective gene is expressed in the kinds of fetal cells collected.

Consider, for example, *Tay-Sachs disease*, which results from a defect in the enzyme *hexosaminidase* A. Because of the defective enzyme, a substance known as *ganglioside* progressively accumulates inside nerve cells and eventually results in blindness, paralysis, and severe mental retardation; affected children usually die before the age of 5. The abnormal enzyme is unusually sensitive to heat and can be detected in skin or blood cells from healthy carriers of the disease, as well as in fetal cells obtained by amniocentesis. The ability to diagnose and abort affected fetuses has greatly helped to reduce the incidence of Tay-Sachs disease, especially in those populations that are at very high risk. As shown in Table 6-4, in the worldwide population about one person in 35,000 is a carrier of Tay-Sachs disease, but among Jews of eastern European origin, one person in 3500 is a carrier.

Of the roughly 200 inborn errors of metabolism for which the exact genetic defect is known, about half can now be detected by prenatal diagnosis. The remaining defects cannot be tested for because the defective genes are not expressed in the kinds of fetal cells obtained by amniocentesis. The abnormal beta chain in sickle-cell disease, for example, is found only in red blood cells, which are not usually obtained by amniocentesis. Until very recently, the prenatal diagnosis of sickle-cell disease depended on the rather risky procedure of obtaining a sample of fetal blood. Fortunately, a better, safer way of diagnosing sickle-cell disease in the fetus has now been devised; this method may open the door to the prenatal diagnosis of many other genetic diseases as well.

Prenatal Diagnosis by DNA Analysis

In 1978, it was reported that DNA from a person with sickle-cell disease yielded a different set of fragments than normal DNA when it was digested by one of a class of enzymes known as *restriction enzymes*. Restriction enzymes cleave the double helix at specific sequences from four to eight bases long. Each restriction enzyme cuts the double helix only where its specific sequence is found (which may be at many places in a given DNA molecule), and nearly 100 different restriction enzymes have been identified so far. As you may recall from the discussion of sickle-cell hemoglobin (hemoglobin S) in Chapter 3, the beta chains of sickle-cell and normal hemoglobin differ in a single base change in the abnormal gene. The base change from A in normal hemoglobin to T in sickle-cell hemoglobin eliminates the specific base sequence recognized by a certain restriction enzyme. Exposing fetal DNA from the chromosomal region that contains the gene for the beta chain to the specific restriction enzyme can therefore reveal the presence of the abnormal gene. DNA that codes for the normal beta chain is digested into two fragments by the restriction enzyme, but the abnormal DNA lacks the specific base sequence and remains intact. DNA extracted from cells obtained by amniocentesis can be analyzed, often without the need for tissue culture. This method of prenatal diagnosis has the great advantage of detecting abnormal genes even in cells that do not usually express the gene under investigation.

DNA analysis (the technique is also known as the analysis of *restriction fragment length polymorphisms*) is most promising in detecting those genetically determined diseases for which the exact biochemical defect remains

TABLE 6–4   SOME COMMON GENETIC DISEASES

Inborn Errors of Metabolism	Approximate Incidence Among Live Births	
Cystic fibrosis (mutated gene unknown)	$\frac{1}{1600}$	Whites
Duchenne muscular dystrophy (mutated gene unknown)	$\frac{1}{3000}$	Boys (X-linked)
Gaucher's disease (defective glucocerebrosidase)	$\frac{1}{2500}$	Ashkenazi Jews
	$\frac{1}{75,000}$	others
Tay-Sachs disease (defective hexosaminidase A)	$\frac{1}{3500}$	Ashkenazi Jews
	$\frac{1}{35,000}$	others
Essential pentosuria (a benign condition)	$\frac{1}{2000}$	Ashkenazi Jews
	$\frac{1}{50,000}$	others
Classic hemophilia (defective clotting factor VIII)	$\frac{1}{10,000}$	Boys (X-linked)
PKU (defective phenylalanine hydroxylase)	$\frac{1}{5000}$	Among Celtic Irish
	$\frac{1}{15,000}$	others
Cystinuria (mutated gene unknown)	$\frac{1}{15,000}$	
Metachromatic leukodystrophy (defective arylsulfatase A)	$\frac{1}{40,000}$	
Galactosemia (defective galactose-1-phosphate uridyl transferase)	$\frac{1}{40,000}$	

Hemoglobinopathies	Approximate Incidence Among Live Births	
Sickle-cell anemia (defective beta-globin chain)	$\frac{1}{400}$	U.S. blacks; in some West African populations, the frequency of heterozygotes is 40 percent
Beta-thalassemia (defective beta-globin chain)	$\frac{1}{400}$	Among some Mediterranean populations

SOURCE: From *Recombinant DNA: A Short Course*, by J. D. Watson, John Tooze, and David T. Kurtz, © 1983. Scientific American Books.

unknown. One such condition is *Huntington's disease*, which results from an autosomal dominant trait that causes cells in specific regions of the brain to gradually die. Affected individuals develop progressively worse involuntary movements and ever-deepening dementia, which end in death 10 to 20 years after the onset of the disease. Huntington's disease is particularly treacherous because its symptoms do not begin to appear until middle age. By that time, an affected individual may already have several children, each of whom has a 50 percent chance of eventually developing the disease. The biochemical defect in Huntington's disease is unknown, but in 1983 it was discovered that when DNA from affected individuals in two different families with a long history of the disorder was digested by certain restriction enzymes, a different set of fragments resulted than when DNA from normal individuals was treated in the same way. Although at present this method is not applicable to other families, there is every reason to believe that in the near future reliable prenatal tests will be developed for Huntington's disease based on a distinctive set of DNA fragments produced by exposing the genetic program of the fetus to the activity of various restriction enzymes. In this case, investigators would also be able to detect individuals who had inherited the abnormal gene but had not yet developed symptoms of the disease. (Surveys of individuals at high risk for Huntington's disease have revealed that at least half of them choose not to know if they were going to develop it.)

In theory, various combinations of restriction enzymes can yield specific sets of DNA fragments for all genetic defects that depend on single abnormal genes. In the future, DNA analysis will probably allow the prenatal diagnosis of certain relatively common inborn errors of metabolism whose exact genetic basis is unknown. A prime target will surely be *cystic fibrosis*, the most common severe inborn error of metabolism in white populations. As shown in Table 6-4, one white person in about every 1600 is a carrier of cystic fibrosis. Affected individuals produce very thick mucus and other body secretions, which makes them especially prone to lung infections, and have an abnormal pancreas, which results in improper digestion. With vigorous, lifelong treatment of the repeated infections, some individuals reach adulthood, but most die of pneumonia during childhood.

Chorionic Villi Sampling as a Possible Replacement for Amniocentesis

Amniocentesis has revolutionized the field of prenatal diagnosis, but it has one major drawback: time. As mentioned earlier, amniocentesis is performed during the 16th to 18th week of pregnancy and the results are usually not available until a month later, which seems like an endless period to the parents. In the United States, about 30,000 women undergo amniocentesis each year, so the total time spent in anxious waiting is an enormous waste. But relief may soon be at hand. In 1983, preliminary results from prenatal diagnosis by a technique known as *chorionic villi sampling* were reported, and they look very promising.

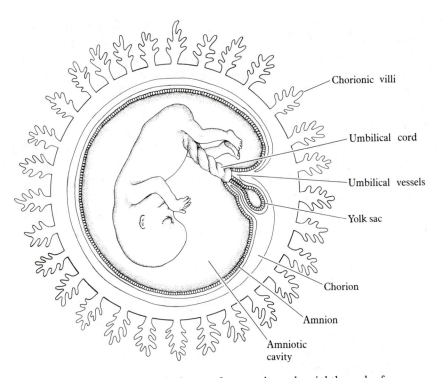

Chorionic villi

Umbilical cord

Umbilical vessels

Yolk sac

Chorion

Amnion

Amniotic
cavity

6–4   Schematic diagram of a human fetus at about the eighth week of pregnancy. In the technique of chorionic villi sampling, a small amount of tissue from the fingerlike chorionic villi is removed by means of a small tube inserted through the vagina and cervix; no amniotic fluid is collected. At this stage of pregnancy, the chorionic villi contain large numbers of rapidly dividing cells. The villi on one side of the fetus gradually enlarge and expand to form the fetal portion of the placenta.

Chorionic villi sampling was first developed in China in 1975 as a means of determining the sex of a fetus. In this technique, a sample is taken of the *chorionic villi*—finger-like projections of the outermost membrane surrounding the fetus (Figure 6-4). Chorionic villi sampling has some decisive advantages over amniocentesis. First, the sample is taken six to nine weeks into the pregnancy. Some investigators consider nine weeks to be the optimal time because, by then, most spontaneous abortions have already occurred. Second, for the pregnant woman the procedure is no more uncomfortable than a routine gynecologic examination, and no anesthesia is required. Third, the results of chorionic villi sampling are available in a few days. Chorionic villi contain large numbers of rapidly dividing cells. Therefore chromosomal studies can be completed within hours after the sample is obtained, and inborn errors of metabolism can usually be detected within a few days. The only shortcoming of chorionic villi sampling compared with amniocentesis is that it cannot detect neural tube defects.

As of 1984, chorionic villi sampling had been used worldwide for nearly 200 women, mostly in England and Italy, and also in the United States. Until many more pregnancies have been completed and the accuracy and safety of the technique have been confirmed, chorionic villi sampling must be considered as a promising but still experimental technique. Preliminary evidence suggests that the inadvertent abortion rate associated with this procedure is not substantially higher than the background rate during early pregnancy. If the procedure proves to be as safe and effective as the initial evidence suggests, chorionic villi sampling may replace amniocentesis as the main technique for the prenatal diagnosis of chromosomal abnormalities and inborn errors of metabolism.

The Dawn of Gene Therapy: The Lesch-Nyhan Syndrome

The Lesch-Nyhan syndrome is a rare inborn error of metabolism that results from a defective X-linked gene. In the most severe cases, the gene product, an enzyme abbreviated HPRT (hypoxanthine-guanine phosphoribosyl transferase) is completely lacking. Affected boys may appear normal as infants, but by the time they are 6 to 8 months old, their nervous system dysfunction begins to become evident. Eventually, involuntary movements begin, kidney stones and severe gout develop, and mental retardation becomes profound. The most distinctive feature of this dreadful affliction, however, is a bizarre form of behavior involving self-destructive biting of the lips and fingers. (This abnormal behavior is discussed further in Chapter 8.)

Several defects in the gene that codes for HPRT have been found in various affected families. Moreover, four different single amino acid substitutions have been described, and in one family a deletion of part of the DNA that codes for the enzyme is responsible for the disease. Decreased or absent HPRT activity can be readily detected in fetal cells obtained by amniocentesis, and prenatal diagnosis of the disease by means of DNA analysis has also been recently developed.

In 1983, two different teams of researchers succeeded in isolating the normal gene for HPRT. The normal gene was spliced into bacterial plasmids—closed circles of double-stranded DNA that replicate independently of the DNA in chromosomes. As illustrated in Figure 6-5, the plasmids containing the spliced-in gene for HPRT were next duplicated many times inside bacterial cells, extracted, and finally introduced into HPRT-deficient cells obtained from a patient with Lesch-Nyhan syndrome and maintained in tissue culture. The presence of the genetically engineered plasmids in the enzyme-deficient cells restored HPRT activity to apparently normal levels. Furthermore, one of the research teams also introduced a certain kind of retrovirus into HPRT-deficient rodent cells along with the plasmid containing the gene for HPRT. From some of the cells, they recovered viruses that had apparently "picked up" the gene for normal HPRT and had incorporated it into their

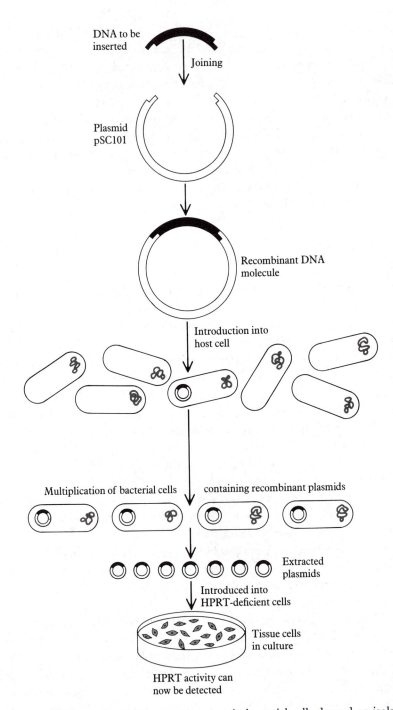

6–5   Techniques for cloning human genes in bacterial cells depend on isolating the gene in question and then splicing it into a plasmid—a self-replicating, circular bit of DNA not associated with the bacterial chromosome. When plasmids are replicated, so is the spliced-in gene, and large numbers of plasmids containing this gene can be produced. (This process is known as *gene cloning*.) The replicated plasmids are then removed from the bacterial cells and introduced into human cells maintained in tissue culture. The plasmids are either directly injected into the cells or enter them by first being incorporated into the genetic material of certain kinds of viruses. (From James D. Watson, John Tooze, and David T. Kurtz; *Recombinant DNA: A Short Course*, copyright © 1983. Scientific American Books.)

genetic programs. In other words, they had produced infectious particles that restore HPRT activity to HPRT-deficient cells.

Although there is probably a long way to go before attempts will be made to introduce the gene for HPRT directly into the cells of patients with Lesch-Nyhan syndrome, experiments like those just described have clearly opened up the possibility of gene therapy—treatment of genetic diseases by the addition of normal DNA to overcome specific defects in the genetic program. Genetically engineered DNA has already been introduced into the genetic programs of several laboratory animals, including mice and fruit flies. Before long, it may be possible to treat certain inborn errors of metabolism by gene therapy, but this is an area of research that clearly calls for caution. *What we are confronting for the first time is the ability to directly alter the genetic programs of human beings.* This awesome ability carries with it equally awesome responsibility. Because of the ethical issues involved, gene therapy may never become widespread or popular, but its possibility and implications are worthy of careful consideration.

Summary

In the United States, there is a 4 percent chance that any newborn will have a birth defect. Parents and other family members usually seek professional counseling concerning the risk of recurrence following the birth of an affected child. Genetic counseling depends on the accurate diagnosis of the abnormality and on the precise understanding of its genetic basis.

Structural malformations are observed in two or three out of every 100 births in the United States, and depend on many genes and certain environmental factors. The exact genetic basis is not known for most kinds of structural malformations, and genetic counseling for these disorders relies on statistics concerning observed recurrence rates among the relatives of affected individuals.

Neural tube defects include anencephaly and spina bifida, both of which can usually be diagnosed by ultrasound imaging of the fetus during the second trimester of pregnancy. The amniotic fluid surrounding fetuses with neural tube defects has elevated levels of alpha-fetoprotein, as does the blood of women carrying affected fetuses. Therefore, measurement of the alpha-fetoprotein level can serve as a valuable screening test for these disorders.

Amniocentesis is usually performed in the 14th to 16th week of pregnancy, and the results are usually not available for about a month. Only about 10 percent of women who are at substantial risk for having a genetically defective child undergo amniocentesis. This technique is very accurate in the prenatal diagnosis of chromosomal abnormalities and in the detection of certain inborn errors of metabolism. Prenatal diagnosis of inborn errors by DNA analysis (restriction fragment length polymorphisms) has increased the number of disorders that can be detected. The DNA from normal and affected individuals yields a distinctive set of fragments when exposed to specific restriction enzymes.

Chorionic villi sampling is probably as accurate as amniocentesis in detecting chromosomal abnormalities and inborn errors of metabolism; it is usually performed in the sixth to ninth week of pregnancy. The results are ready in a few days. Chorionic villi sampling may one day largely replace amniocentesis, but it cannot detect neural tube defects.

It will soon be possible to treat genetic diseases by introducing normal DNA into the defective DNA of affected individuals. The ability to alter the genetic programs of human beings raises serious ethical questions that must be addressed.

Suggested Readings

"Prenatal Diagnosis of Genetic Disease," by Theodore Friedman. *Scientific American*, Nov. 1971, Offprint 1234. A discussion of amniocentesis and other diagnostic techniques that have both genetic and social aspects.

"The Manipulation of Genes," by Stanley N. Cohen. *Scientific American*, July 1975, Offprint 1324. Discusses some of the molecular details and social implications of the new technique of genetic engineering.

"Genetic Load," by Christopher Wills. *Scientific American*, March 1970, Offprint 1172. How the accumulated mutations of any species are usually detrimental but at the same time may be a priceless genetic resource.

"Prenatal Diagnosis of Genetic Defects," by Renata Laxova. *Postgraduate Medicine*, 3 March 1979. Provides statistics on the reasons for which amniocentesis is performed and discusses how possible screening programs for genetic defects might be organized.

"Genetic Counseling," by Mark Degnan et al. *American Family Physician*, July 1975. Includes a sensitive discussion of the psychological effects on the parents of a genetically abnormal child.

"The Revolution in Fetal Health Care," by Fran Pollner. *Medical World News*, 12 Sept. 1983. Nontechnical review article that provides information about chorionic villi sampling and other means of prenatal diagnosis.

"Genetic Amniocentesis," by Fritz Fuchs. *Scientific American*, June 1980. An update on the most important means of obtaining fetal cells for prenatal analysis.

"Towards Gene Therapy: Lesch-Nyhan Syndrome," by Julie Ann Miller. *Science News*, 6 Aug. 1983. Provides some details on the recent experiments mentioned in this book.

"Huntington's Disease Gene Located," by Gina Kolata. *Science*, 25 Nov. 1983. How restriction enzymes were employed in the recent isolation of the gene for Huntington's disease.

Clinical Symposia: Understanding Inherited Metabolic Disease, by William L. Nyhan. CIBA Pharmaceutical Company, 1980. A relatively nontechnical presentation of the clinical features of some of the most common inborn errors of metabolism.

Chapter 7

Genes in the Human Population

Our precious planet is home to nearly 6 billion human beings, yet no two individuals, not even identical twins, are exactly alike. Human variation, like that of all other living things, depends both on genes and on environmental factors. The most obvious differences between persons are features of the body surface and body proportions. Each of us has a unique face and absolutely distinctive fingerprints, and some of us are tall or short, black or white, fat or thin, have curly or straight hair, broad or narrow noses, and so on. At the biochemical level, human beings are as diverse as they are on the physical level. For example, each of us belongs to one of the more than 60 different blood groups, can have one or more of several alternative forms of enzymes and other proteins, and has surface features on our body cells that are nearly as distinctive as our faces and fingerprints.

In 1930, the mathematician Ronald Fisher (Figure 7-1) published a book entitled *The Genetical Theory of Natural Selection*, in which he demonstrated that the diversity of genetically determined traits in a given population can be directly related to fitness for survival. Fisher's equations indicated that the genetic diversity observed in most populations exists mostly because a high degree of variability allows for more possibilities of adaption to environmental conditions and, therefore, for successful reproduction and the contribution of offspring to the next generation. Because offspring usually inherit many of the characteristics that had favored the survival of their parents, adaptive traits eventually come to predominate in the population over time. The differential reproduction of organisms with advantageous heritable characteristics is known as *natural selection.*

In recent years, it has been discovered that individual animals of the same species, including human beings, are much more variable than Fisher or anyone else realized only a few decades ago, and that many genetically influenced traits confer no obvious reproductive advantage on the individuals possessing them. In this chapter, our main concern is to explain how the frequencies of various genes observed in different segments of the human population can be influenced, not only by natural selection but also by isolation, chance, and other factors. We shall also consider the concept of biological race and the possible reasons for the frequencies and distribution of certain traits in the present population. As mentioned in Chapter 3, spon-

7–1   Sir Ronald Fisher was one of the first persons to formulate equations that describe how natural selection can influence a population's genetic composition. (Courtesy of Godfrey Argent.)

This portrait of men of various human races is based on original photographs taken by Professor Carleton Coon, who kindly granted permission to have this drawing rendered. The groups represented are (left to right): front row, Armenian, Formosan, Bavarian, Veddoid; middle row, Dinaric, Singhalese, Arab; back row, Negrito, Korean, Swedish, Moroccan. Photos of women of various races are found in Figure 7-11.

taneous changes in DNA, known as *mutations*, are ultimately responsible for all genetically determined diversity. In concluding this chapter, the role of mutations in maintaining diversity and the effects of various environmental factors on the rate of mutation will be discussed.

The Polymorphic Nature of Many Human Proteins

As discussed in Chapter 3, protein molecules consist of long chains of amino acids linked in tandem by chemical bonds. There are about 20 kinds of amino acids, and the properties of a particular protein depend above all on the sequence of amino acids in its chain. What determines the sequence of amino acids in a protein is the sequence of the four bases along the particular stretch of the DNA molecule that codes for the protein in question.

A relatively easy and sensitive way to determine whether protein molecules differ from one another is to observe their patterns of movement in an electric field. Most protein molecules, or regions of protein molecules, have either a positive or a negative charge. The distribution of the charge depends on which amino acids are present in the protein and on how they are arranged. Proteins that differ from one another in their amino acid sequences usually have different electrical properties and therefore show different patterns of movement in an electric field. In general, these patterns reflect biochemical differences that are genetically determined. As you may recall, a difference of only one amino acid out of a total of 287 in the protein portion of the hemoglobin molecule can be easily detected by this method (see Figure 3-14).

Among the proteins coded for by the thousands of structural genes in the human genetic program, only a small fraction have had their exact amino acid sequences worked out. Of the more than 100 enzymes for which the amino acid sequence is known, about 25 percent of these enzymes exist in several different forms, each of which has a unique amino acid sequence and unique properties. Populations in which several alternative forms of a gene or gene product are regularly encountered are said to be *polymorphic* (having many forms) for that gene or gene product. Table 7-1 lists the frequencies of the most highly polymorphic enzymes in the British population; each person has one or two forms of each enzyme. Even though alternative forms of the same enzyme may be regularly found in a given population, there is little or no evidence of how or whether natural selection maintains them because of differential reproduction. It may be true that most enzyme polymorphisms, which confer no obvious advantage or disadvantage on the person possessing them, are not affected by natural selection. At present, we do not know. In addition to widespread enzyme polymorphisms that are regularly encountered in the worldwide human population, rare variants of practically all known enzymes also occur. It is estimated that about one or two persons per 1000 carry a rare, apparently nondetrimental variant of some enzyme.

We have already discussed how one protein polymorphism—that of the beta chain of sickle-cell hemoglobin (hemoglobin S) and normal hemoglobin (hemoglobin A)—can be explained by natural selection. Hemoglobin is found inside red blood cells, which are the primary targets for invasion by

TABLE 7–1 FREQUENCIES OF VARIOUS POLYMORPHIC VARIANTS OF ENZYMES IN THE BRITISH POPULATION

Enzyme	Form					
	1	2	3	$\frac{1}{2}$	$\frac{2}{3}$	$\frac{1}{3}$
Red-cell acid phosphatase	.13	.36	0	.43	.05	.03
Phosphoglucomutase 1	.59	.06	——	.35	——	——
Phosphoglucomutase 3	.55	.07	——	.38	——	——
Placental alkaline phosphatase	.41	.07	.01	.35	.05	.12
Peptidase A	.58	.06	——	.36	——	——
Adenylate kinase	.90	.01	——	.09	——	——
Adenosine deaminase	.88	.01	——	.11	——	——
Alcohol dehydrogenase 2	.94	——	——	.06	——	——
Alcohol dehydrogenase 3	.36	.16	——	.48	——	——
Glutamate-pyruvate transaminase	.25	.25	——	.50	——	——
Esterase D	.82	.01	——	.17	——	——
Malic enzyme	.48	.09	——	.43	——	——
Phosphoglycolate phosphatase	.68	.03		.29	——	——
Glyoxylase 1	.30	.21		.49	——	——
Diaphorase 3	.58	.05		.36	——	——

SOURCE: From *The Principles of Human Biochemical Genetics,* 3d ed., by H. Harris. North-Holland 1980.

the parasite that causes malaria. Sickle-cell hemoglobin somehow inhibits the reproduction of the parasite, and persons with hemoglobin S in their red cells are therefore more resistant to malaria than those whose red cells contain only hemoglobin A. Because being heterozygous for the gene for sickle-cell hemoglobin confers resistance to malaria, heterozygotes survive and reproduce at a higher rate than individuals who lack the gene. This accounts for the worldwide distribution of the gene for sickle-cell hemoglobin in native populations. As shown in Figure 3-17, distribution of the gene for hemoglobin S is nearly the same as that of the parasite that causes malaria. In those populations in which hemoglobin S is regularly encountered, from 25 to 30 percent of the individuals have both hemoglobins S and A in their red cells. Where malaria does not occur, hemoglobin S is rare. Furthermore, when native individuals leave areas where malaria is widespread and settle in nonmalarious areas, the frequency of the gene for hemoglobin S among their offspring decreases. Among American blacks, most of whom are the descendants of blacks from West Africa, where malaria had long been endemic, the incidence of heterozygotes for hemoglobin S is from 8 to 10 percent.

There is evidence suggesting that the gene that codes for the abnormal beta chain in beta-thalassemia may also be maintained by natural selection because its abnormal gene product confers resistance to malaria. In this case, the abnormal gene is regularly found in the genetic programs of persons living in the Mediterranean region, where malaria has now been practically eliminated. Because being heterozygous for beta-thalassemia no longer provides a reproductive advantage, the frequency of the abnormal gene is expected to decrease in future generations.

We now turn to a well-documented, though incompletely understood, genetic polymorphism on which reliable data are available from nearly all human populations. Perhaps the most extensively studied human polymorphism is the ABO blood group.

The Worldwide Frequencies and Distribution of the ABO Blood Group

The genes that determine to which ABO blood group an individual belongs come in three alternative forms, or alleles, that can by symbolized I^A, I^B, and I^O. Each human being has one of the three alleles at the same location on each copy of chromosome 9. The patterns of inheritance shown by these three alleles and how they determine to which ABO blood group a person belongs are discussed in Chapter 1. The worldwide frequencies of these three alleles in the human population are known with considerable accuracy. As determined by medical and anthropological surveys, the worldwide frequencies are: I^O, 62 percent; I^A, 22 percent; and I^B, 16 percent. In most European populations, about 45 percent of the people are of type O ($I^O I^O$), 35 percent are of type A ($I^A I^A$ or $I^A I^O$), 15 percent are of type B ($I^B I^B$ or $I^B I^O$), and 5 percent are of type AB ($I^A I^B$).

7–2   The distribution of the allele I^O in aboriginal populations around the world. (After Mourant et al., *The ABO Blood Groups*. Copyright © 1958 by Blackwell Scientific Publications.)

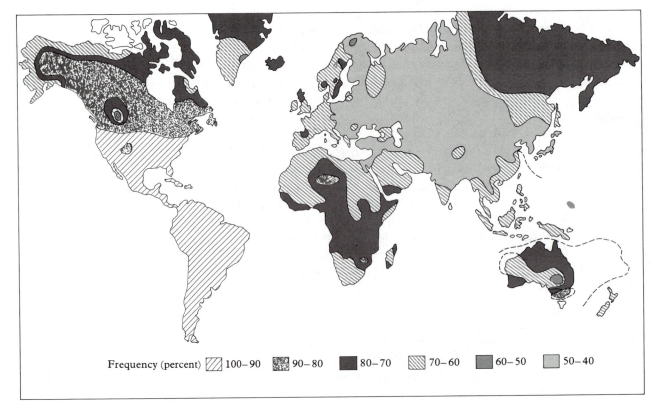

Frequency (percent)   100–90   90–80   80–70   70–60   60–50   50–40

The alleles for the ABO blood group are by no means distributed equally throughout the world. Figure 7-2 shows the distribution of the allele I^O in aboriginal populations around the world. Notice that North and South American Indians have very high frequencies of the allele I^O. In fact, in some areas of these two continents, the aboriginal peoples have the allele I^O almost to the exclusion of the other two. On the other hand, the distribution of the allele I^B is almost opposite to that of I^O, as shown in Figure 7-3. I^B is very common in Central Asia but rare among native North and South Americans.

How can we account for these definite patterns in the worldwide distribution of the alleles that determine to which ABO blood group a person belongs? Perhaps the most likely explanation is that the distribution of ABO blood groups is one effect of human migrations, most of which took place in prehistoric times. Human beings as we know them today probably first evolved in Africa or Asia. Initially human populations must have been rather small, but with their characteristic resourcefulness, people soon invented food cultivation and then increased their numbers and extended their range, not only to the entire African and Asian continents, but to the rest of the Old World as well. Eventually, people made their ways to the Americas, the islands of the Pacific, and Australia. These migrations could easily have resulted in the gradual emergence of the present-day distribution of the ABO blood groups.

One example of how the distribution of ABO blood groups can be influenced by human migrations is shown in Figure 7-4, which is a close-up map of the distribution of the allele I^B in Eurasia. In general, the frequency of I^B

7–3   The distribution of the allele I^B in aboriginal populations around the world. (After Mourant et al., *The ABO Blood Groups.* Copyright © 1958 by Blackwell Scientific Publications.)

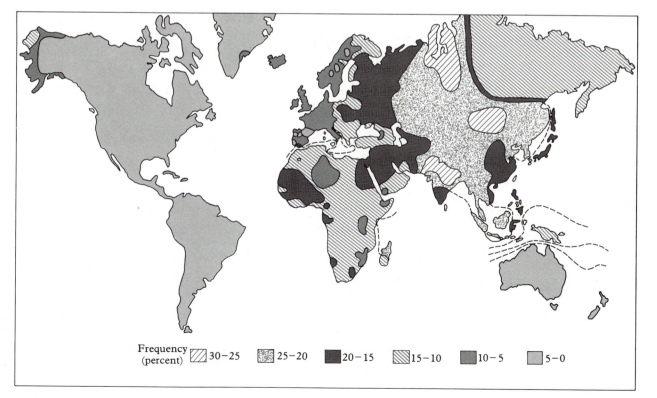

Frequency (percent) 🔲 30–25   🔲 25–20   ◼ 20–15   🔲 15–10   🔲 10–5   🔲 5–0

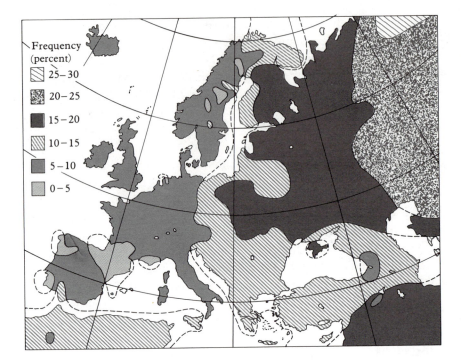

7–4   The distribution of the allele I^B in Eurasia. Note the relatively low frequencies of I^B near the Pyreness and Caucasus mountains. (After Mourant et al., *The ABO Blood Groups.* Copyright © 1958 by Blackwell Scientific Publications.)

Frequency (percent)
- 25–30
- 20–25
- 15–20
- 10–15
- 5–10
- 0–5

shows a steady decrease from Central Asia toward Western Europe. This pattern has been attributed to the effects of Mongol invasions of Europe, which continued for about 1000 years and ended about 500 years ago. The Mongols from the East may have had proportionately higher concentrations of I^B than their Western counterparts, and as the former moved westward they undoubtedly spread not only their culture but their genes as well. The distribution of group B in Western Europe could be explained by assuming that before the Mongol invasions most Western Europeans did not have the allele I^B. A comparison of Figures 7-2 and 7-4 suggests that early Western Europeans probably had a preponderance of the allele I^O instead, as evidence by the higher frequency of I^O in extreme Western Europe today. (The very low concentrations of I^B that are found today in the area of the Pyrenees Mountains may have resulted because some of the original Europeans fled to the mountains to evade the Mongol hordes.)

Overall, there is little doubt that human migrations have strongly influenced the present distribution of the ABO blood group. But what about the effects of natural selection? The difficulty in trying to assess its importance in determining the frequencies and distribution of the ABO blood group is that we have no decisive evidence about whether or not any one of the three alleles is directly influenced by natural selection. That is, human beings of one ABO blood group enjoy no obvious reproductive advantage over those of any other ABO group.

But this is not to say that statistical correlations between ABO blood groups and various diseases do not exist. In fact, based on data from Great Britain in the 1950s, there is evidence that members of group O are about 40 percent more likely to develop a duodenal ulcer than members of the other two blood groups. What is not clear, however, is how or whether duodenal ulcers are related to natural selection. If the two are related, then we would expect persons who have ulcers to produce either more or fewer offspring than those who do not, but there is no evidence that this happens. It seems unlikely that natural selection has influenced the worldwide distribution and frequencies of the ABO blood groups simply because of the statistical correlation between group O and ulcers.

On the other hand, it has been suggested that natural selection may have influenced the present-day distribution and frequencies of the ABO blood groups because persons of different groups may be more resistant to certain infectious diseases. For example, persons of groups B and O have natural antibodies against group A in their bloodstreams. Antibody against group A may also be effective against the virus that causes smallpox. Thus, members of groups B and O may be more resistant to smallpox than those of group A, and the present-day frequencies and distribution of the ABO groups could reflect devastating epidemics of smallpox or other infectious diseases in the past.

Natural selection could also affect the ABO groups because of antibody-dependent incompatibilities between a mother and her developing fetus. For example, if a woman is type O and her fetus is type A, there may be difficulty because the mother has natural antibodies against the red cells of the fetus. But data about the actual outcome of ABO-incompatible pregnancies—although abundant—are inconclusive, and some of them are contradictory. For now, we don't know whether natural selection significantly affects the frequencies of ABO blood groups because of antibody-dependent incompatibilities between mother and fetus.

To sum up, natural selection has undoubtedly played a role in determining the frequencies and distribution of the ABO blood groups, but it has probably not been the most important factor. Rather, the present-day frequencies and distribution of the ABO blood groups in human population can perhaps best be explained by prehistoric migrations and the effects of chance.

Highly Polymorphic Nature of the Genes of the Major Histocompatibility Complex

Grafts of tissue or transplants of organs from one mammal to another are usually recognized and destroyed by the host's immune system in a matter of weeks or months. The recognition of "self" and the rejection of "nonself" are complex processes that depend largely on the presence of a distinctive set of protein molecules attached to the surfaces of nearly all body cells. Surface proteins are particularly abundant on the white blood cells known as *lymphocytes*, which are important mediators in the rejection of transplanted cells, tissues, or organs. Most of the surface proteins are coded for by a cluster of genes known as the *major histocompatibility complex* (MHC), which in humans is located on chromosome 6.

In the past 25 years, huge amounts of data have been gathered concerning the variability of genes in the MHC and their distribution in various populations. The cell surface proteins coded for by the genes of the MHC are

TABLE 7–2   FREQUENCIES OF DIFFERENT TYPES OF HLA ANTIGENS
IN SAMPLES FROM THREE POPULATIONS

	European Caucasoids	African Blacks	Japanese		European Caucasoids	African Blacks	Japanese
HLA-A Antigen				**HLA-B** Antigen			
A1	.16	.04	.01	B5	.06	.03	.21
A2	.27	.09	.25	B7	.10	.07	.07
A3	.13	.06	.007	B8	.09	.07	.002
A23	.02	.11	—	B12	.17	.13	.07
A24	.09	.02	.37	B13	.03	.02	.008
A25	.02	.04	—	B14	.02	.04	.005
A26	.04	.05	.13	B18	.06	.02	—
A11	.05	—	.07	B27	.05	—	.003
A28	.04	.09	—	B15	.05	.03	.09
Aw29	.06	.06	.002	Bw38	.02	—	.02
Aw30	.04	.22	.005	Bw39	.04	.02	.05
Aw31	.02	.04	.087	B17	.06	.16	.006
Aw32	.03	.02	.005	Bw21	.02	.02	.02
Aw33	.007	.01	.02	Bw22	.04	—	.07
Aw43	—	.04	—	Bw35	.10	.07	.09
Blank	.02	.11	.04	B37	.01	—	.008
				B40	.08	.02	.22
				Bw41	—	.02	—
				Bw42	—	.12	—
				Blank	.04	.18	.08
HLA-C Antigen				**HLA-DR** Antigen			
Cw1	.05	—	.11	DRw1	.06	—	.05
Cw2	.05	.11	.01	DRw2	.11	.09	.17
Cw3	.09	.06	.16	DRw3	.09	.12	—
Cw4	.13	.14	.04	DRw4	.08	.04	.14
Cw5	.08	.01	.01	DRw5	.15	.07	.05
Cw6	.13	.18	.02	DRw6	.09	.10	.07
Blank	.47	.50	.53	DRw7	.16	.07	—
				W1A8	.06	.07	.07
				Blank	.21	.45	.45

SOURCE: From *The Principles of Human Biochemical Genetics,* 3d ed.; by H. Harris.
North Holland 1980.

most easily detected on the surfaces of lymphocytes and are known as *human lymphocyte antigens* (HLAs). (An *antigen* is a substance that elicits the production of a specific antibody, and the cell surface proteins are excellent antigens.) There are four main categories of HLAs, designated A, B, C, and DR. Each person has two copies of an allele in each category, and each category is highly polymorphic. Fifteen varieties of the A allele have been identified in European populations, 18 varieties of the B allele, 7 varieties of the C allele, and 9 varieties of the DR allele. As shown in Table 7-2, none of the alleles in the MHC is very common, and the number of possible combinations is enormous. Among the MHC genes discovered so far, about *25 million* different combinations of alleles are possible. It is no wonder that most attempts to transplant tissues or organs between two unrelated persons fail because of rejection. The more nearly identical the combination of alleles in the donor and receiver, the better the chance that the graft or transplant will be accepted. Transplanted kidneys from identical twins, whose MHC genes are the same, survive indefinitely, as do those transplanted between donors and receivers who by chance happen to have nearly identical MHC genes.

The MHC alleles are the most polymorphic set of alleles in the human genetic program. Why are they so amazingly diverse? Nobody knows for sure, but the extreme diversity of the MHC alleles may help to provide protection against various kinds of infectious diseases. In order for many bacteria and viruses to colonize human cells, they must first become attached to specific cell surface proteins known as *receptors*. The diversity of surface proteins may make the receptors harder for the virus or bacterium to find or may provide protection in some other way. Another reasonable speculation is that the diversity of MHC alleles is involved in the recognition of cancer cells as nonself and in their subsequent elimination. Certain MHC alleles has been regularly associated with the presence of, or a tendency to develop, a wide range of human diseases. The most striking example is *ankylosing spondylitis*, a severe form of arthritis of the spine. Although the incidence of the MHC allele designated HLA-B27 in white European populations is only 5 percent, 90 percent of persons who have ankylosing spondylitis have the allele. Many other less striking associations between diseases and MHC genes have been established, and although their significance is not clear, this information can be helpful in identifying certain hard-to-diagnose conditions and in detecting apparently normal individuals who may be at high risk.

As shown in Table 7-2, the frequencies of the alleles of the MHC vary from one human population to another. Thus, the incidence of HLA-B27 is 5 percent in European whites, 0.3 percent in the Japanese, and zero in African blacks. More extreme variations in frequency may occur among smaller populations that are relatively isolated from neighboring ones. for example, 18 percent of Pima Indians in the United States and 50 percent of Haida Indians in Canada have the allele HLA-B27. These high percentages of HLA-B27 among small, relatively isolated populations probably resulted from the effects of chance. That chance can play a decisive role in determining the genetic composition of certain human populations is well known. One example of how this occurs is provided by the Dunkers, a group who migrated from Germany to Pennsylvania in the early eighteenth century.

Genetic Drift

In populations made up of large numbers of persons who mate with one another at random, the frequencies of those genes not obviously influenced by natural selection tend to remain about the same from one generation to the next. This is because the frequencies of such genes are determined mainly by chance, and in large populations chance fluctuations in gene frequencies tend to balance one another. But in relatively small populations this does not necessarily happen, and significant changes in the genetic composition of the population can occur by chance alone. This chance variation in gene fre-

7–5   The Old Order Dunkers of Franklin County, Pennsylvania. Although they seldom marry outside the sect and their attire differs from that of their neighbors, their customs are not otherwise unusual. (From "The Genetics of the Dunkers," by H. Bentley Glass. Copyright © 1953 by Scientific American, Inc. All rights reserved.)

quencies from one generation to another is known as *genetic drift,* and in general the smaller the population the greater the genetic drift can be.

An ideal population in which to examine the effects of genetic drift is the devout Protestant religious sect known as Dunkers (Figure 7-5). Between 1719 and 1729, 50 families of Dunkers emigrated from the German Rhineland to Pennsylvania and thereby completely transplanted the sect to the New World. To marry outside the church is considered a grave offense, and a Dunker who does so must either withdraw voluntarily from the community or be expelled from it.

During their first 100 years in Pennsylvania the Dunkers doubled in number, and almost all of them could trace their ancestry to the original 50 families. Then in 1882 the Dunker church underwent schism and a progressive group separated from the old order. At that time the Old Order Dunkers numbered about 3000, and this number has not changed much to the present day. One of the original Dunker communities, in Franklin County, Pennsylvania, remained with the old order, and in the years since 1882 its size has also changed very little. When the Franklin County group of Dunkers was studied in the early 1950s, the population was about 300 and its size had been nearly the same for several generations. The Franklin County Dunkers were thus an almost ideal population in which to look for the effects of genetic drift.

The effects of genetic drift on the Franklin County Dunkers can be revealed by a comparison of the frequencies of various genes among the Dunkers, their West German forebears, and their present-day American neighbors. For example, these are the frequencies of the ABO blood group alleles I^O, I^A, and I^B among the Dunkers, West Germans, and Americans:

	I^O	I^A	I^B
Dunkers	60%	38%	2%
West Germans	64%	29%	7%
Americans	70%	26%	4%

Notice that the frequencies of these alleles among the Dunkers are not the same as those of West Germans or Americans; nor do they lie between the two. This is to be expected if genetic drift is at work and the frequencies of the alleles are determined largely by chance.

Blood groups are not the only characteristic of the Dunkers that shows the effects of genetic drift. There are clear-cut differences between the Dunkers and surrounding American communities in the frequencies of several other apparently nonadaptive but genetically determined traits. Thus, compared with their neighbors, fewer Dunkers have hair on the middle segment of one or more fingers, fewer are able to bend the end of the thumb backward to form an angle of more than 50°, and fewer have earlobes attached to the side of their heads rather than hanging free (Figure 7-6). The best explanation for these observations is that the frequencies among the Dunkers are the result of genetic drift.

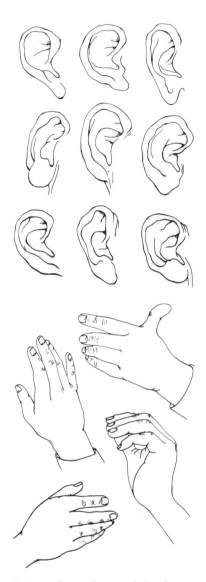

7–6   Some characteristics that show the effects of genetic drift in the Dunkers. Earlobes can be attached or they can hang free; those of most Dunkers hang free. Dunkers have a high incidence of hitchhiker's thumb (the ability to bend the thumb backwards at an angle of more than 50 degrees), and, as compared with the general population, fewer Dunkers have hair on the middle segment of one or more fingers. (From "The Genetics of the Dunkers," by H. Bentley Glass. Copyright © 1953 by Scientific American, Inc. All rights reserved.)

The Dunkers are not the only human population in which genetic drift has been detected. Marked variations in the frequencies of certain alleles from those of neighboring populations have also been reported in other populations, including aboriginal Australians, Eskimos, Italians in isolated villages, North American Indians, and religious sects in Montana. All of these groups have in common two important features that enable genetic drift to strongly influence their genetic constitutions. First, the populations are relatively small, and second, the groups are isolated from neighboring populations either by physical barriers or by cultural rules.

Although the effects of genetic drift are most pronounced in the smallest populations, genetic drift can also influence the genetic composition of larger populations. It has been estimated that genetic drift can strongly influence the frequencies of apparently nonadaptive traits in human populations if the parents in any generation number a few hundred individuals or fewer. This is important because before the invention of food cultivation more than 10,000 years ago, most human populations were probably within this size range and were therefore small enough to be strongly influenced by genetic drift. In fact, many inherited differences between human beings belonging to different races may have become established thousands of years ago when people lived in small groups that were physically and reproductively isolated from one another. Let us discuss the concept of *biological race* and its relation to genetic drift and to natural selection.

The Biology of Human Races

The species to which all living people belong, known as *Homo sapiens*, originated in either Africa or Asia and then spread outward. From the start, our species has been endowed with dextrous, tool-making hands that are controlled by the most complicated organ that evolution has produced so far—the human brain. The combination of human hand and human brain can be thought of as an adaptation that has allowed our species to virtually cover the surface of the land. No other animal species is as widely distributed as the human species, with the possible exception of species such as houseflies, body lice, and mice, which directly benefit from human activities and which have therefore followed people in their migrations over the continents.

When widespread species are examined over their full geographic range, it is often found that populations in different places look slightly different. For example, song sparrows from New York and Oregon can easily be told apart, as can zebras from different parts of Africa (Figure 7-7). As we all know, the human species is no exception when it comes to geographic variation. Thus, people from Tokyo, Copenhagen, Bombay, and Nairobi can be told apart as easily as the song sparrow and zebras from different locations. What accounts for the differences in appearance between populations of the same species?

These variations are the result of the interaction of the genes of a population with the environment. The environment determines which genes from among the total range of genes in the species' DNA will be expressed and to what degree. Nonetheless, most populations of widespread species look

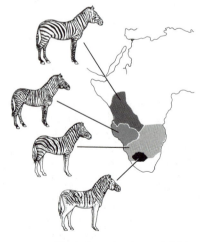

7–7   Variation of stripes of African zebras in different geographic regions. (After Cabrera, *Journal of Mammology*, vol. 17, 1936.)

THE BIOLOGY OF HUMAN RACES

slightly different from one another because of genetic differences between the groups. These genetic differences arise because populations that are separated from one another by long distances or other barriers cannot interbreed. Whenever local populations of a given species are isolated from one another by a barrier, genetic differences accumulate because of the effects of natural selection, mutation, and, if the populations are small enough, genetic drift. Distinct local breeding groups may thus arise in different areas of a widespread species' range. These distinct local breeding groups within a particular species are known as *races*.

There is no doubt that human races originated in the same way as other animal races. About 40,000 years ago, most human beings probably lived in small tribes that were relatively isolated from one another by distance and custom. This isolation resulted in chance differences in genetic constitution from one group to another. During this time of relative isolation among early human groups, some of the genetic traits that characterize modern races probably became established by genetic drift.

But chance was not the only factor favoring the development of genetic differences between isolated groups of ancient people. Surely natural selection must also have played a role in the evolution of human races. As our ancestors increased in number and extended their range, different groups found themselves in very different environments. You will recall that because of natural selection, organisms tend to acquire traits that allow them to adapt closely to the local environment. Thus it seems likely that at least some racial characteristics became established because they were advantageous under certain environmental conditions.

Perhaps the most apparent human racial characteristic is the color of a person's skin. At least 36 shades of human skin color, ranging from jet black to almost white, have been described. As shown in Figure 7-8, there is undoubtedly a connection between skin color and the intensity of sunlight (especially its ultraviolet component). In general, the most darkly pigmented people live closest to the equator and are exposed to the greatest concentration of sunlight. In some parts of the world, especially Africa, skin color shows a steady gradation from darker to lighter the farther away from the equator the population lives. In other locations, changes in pigment with latitude are not so clear-cut, and the original pattern has been blurred and distorted by human mobility and racial mixing in recent years.

Because our species almost certainly evolved first in the tropics, it is likely that all human beings were at first darkly pigmented. During the early stages of human evolution, natural selection may have favored dark skin, not only because of the protection it provides against ultraviolet radiation, but also because a dark color may have provided better camouflage for our ancestors, who must have been prey for some of the large carnivores of the day. As people migrated from the tropics to more northerly or southerly regions, they may have progressively lost most of their skin pigmentation.

You may be wondering why people possess some genetically determined traits, such as skin color, in seemingly endless varieties that differ slightly, whereas they either have or do not have others, such as the ability to synthesize hemoglobin S. The main reason is that the former are under the

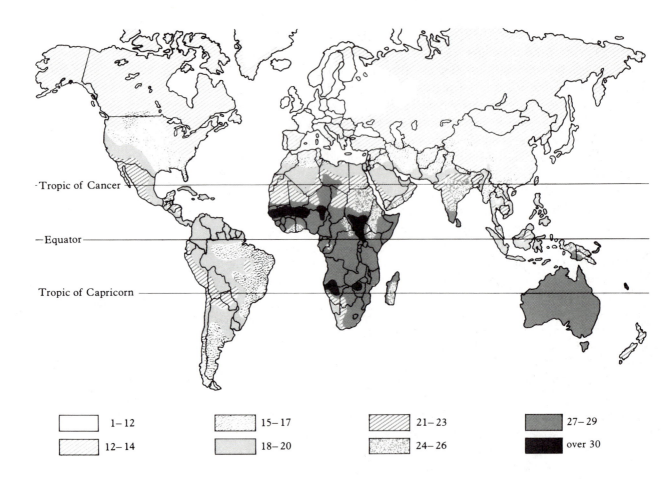

	1–12		15–17		21–23		27–29
	12–14		18–20		24–26		over 30

7–8   The distribution of human skin color before Columbus's first voyage to the New World in 1492 A.D. The values increase with darker skin color. (After R. Baisutti, *Razze e Popoli della Terra*, Torino: UTET, 1951.)

influence of several pairs of alleles, whereas the latter depend on a single pair. Thus, the regular variation in skin color among human populations around the world is probably the result of the interaction of at least four alleles, as shown in Figure 7-9. That these alleles are (or were) retained in different frequencies in different human populations can probably be attributed to the effects of natural selection. (Human characteristics that depend on the interactions of several alleles are discussed in the following chapter.)

Another characteristic that can show gradual variation with latitude, that depends on several pairs of alleles, and that is influenced by natural selection is body build. In general, warm-blooded animals that live in hot equatorial climates are smaller, have longer arms and legs, and have larger ratios of surface area to body weight than warm-blooded animals that live farther north or south. Most of these tendencies are not very clear-cut in the human population, and there are many exceptions to the rule. Nonetheless, as shown in Figure 7-10, human adaptation to climate probably accounts for the differences in body build that allow tall, slender Africans to dissipate more unneeded body heat than Eskimos, who because of where they live must conserve as much body heat as possible.

It is likely that other body features, such as the size of the nose, the color and form of the hair, and the shape of the eyefolds also became established in certain populations because of natural selection. This is because all of these characteristics are external, as are skin color and body build. We would expect the superficial, readily visible characteristics to experience the effects of natural selection because our body surfaces are the interface between our bodies and the environment. It is therefore no wonder that human body surfaces have been altered by natural selection to best fit the varied environments into which our ancestors migrated (Figure 7-11).

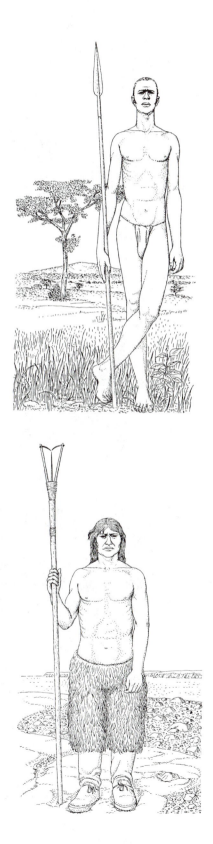

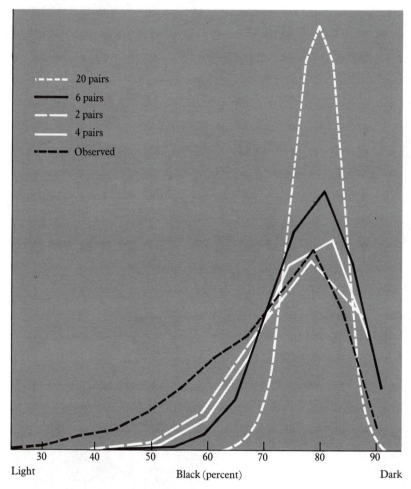

7–9 The distribution of the observed skin color of American blacks and the distribution that would be expected if 2, 4, 6, and 20 pairs of alleles were acting together. (After Curt Stern.)

7–10 The greater body surface of the Nilotic black from the Sudan (top) dissipates unneeded body heat, whereas the proportionately greater bulk of the Eskimo (bottom) conserves body heat. (From "The Distribution of Man", by William W. Howells. Copyright © 1960 by Scientific American, Inc. All rights reserved.)

7–11 These photos of women of various human races reveal superficial differences in the contours of the face that may have evolved under the influence of natural selection on our distant ancestors in various parts of the world. Opposite page, top left, young Negrito woman and her children; top right, a Polynesian woman (probably Tahitian); bottom, Mongolian woman from Inner Mongolia. This page, top left, Singhalese woman; top right, young Swedish woman; bottom Mamayquk Eskimo woman. (Photographs courtesy of the American Museum of Natural History.)

Biochemical Differences Between Human Races

But differences in traits that are neither superficial nor readily visible parallel the superficial differences of human races to only a limited degree. You will recall that among the 100 or so enzymes studied so far, about 25 percent exist in several alternate forms. In recent years it has been possible to compare differences in protein molecules within and between the various races of people. For those protein polymorphisms not obviously influenced by natural selection (and this includes the great majority), the differences in protein molecules between living human races are slight. And protein differences are as slight between whites and African blacks as between whites and Orientals. Moreover, the protein molecules of two whites from opposite ends of Europe differ more than the molecules of two whites from the same isolated European village, but in both differences are about the same as those between the molecules of whites and African blacks or Orientals. Thus, at the biochemical level, the differences between human races are generally much less pronounced than the superficial differences that are apparent on body surfaces. In fact, there is so much biochemical overlap that people cannot be assigned with certainty to a given race solely on the particular alleles contained within their genetic programs. Overall, it is estimated that some 85 percent of human genetically determined variation is between individuals in a given nation, tribe, or other cultural group in which one race usually predominates. The remaining 15 percent of human genetically determined variability is divided evenly between variation between nations of a given race and variation between one race and another.

The Number of Human Races

How many human races are there? That depends on how the term *race* is defined. The best definition is that races are local breeding groups within a particular species. Races are thus defined by genetic relations between populations and not by differences between individuals. But most local breeding groups in any species tend to blend with one another at their geographic boundaries because adjacent races interchange genes when they reproduce. Thus, races are never clear-cut, precisely defined entities. There are undoubtedly thousands of local breeding groups within the human population today that could legitimately be defined as races. The Dunkers are one example, as are relatively isolated populations in Alpine villages, New Guinea, and Australia.

But it is neither practical nor particularly informative to designate every isolated group of people as a separate race. Rather, most of the time the term *race* is used for any relatively isolated local breeding group that is convenient to distinguish for purposes of a given study. In other words, the number of human races described depends largely on the purposes of the describer.

For the most part, the description of human races has been undertaken, appropriately, by anthropologists, but some of the criteria used to distinguish individual races are at best vague. The number of human races currently recognized by different anthropologists using different criteria ranges from zero up, but most estimates fall somewhere in between three and thirty human

TABLE 7–3   VARIOUS HUMAN "RACES" THAT CAN BE IDENTIFIED BY MEANS OF STATISTICAL CORRELATIONS OF THE STRUCTURES OF A VARIETY OF PROTEIN MOLECULES*

CAUCASIANS

Arabs, Armenians, Austrians, Basques, Belgians, Bulgarians, Czechs, Danes, Dutch, Egyptians, English, Estonians, Finns, French, Georgians, Germans, Greeks, Gypsies, Hungarians, Icelanders, Indians (Hindi speaking), Italians, Irani, Norwegians, Oriental Jews, Pakistani (Urduspeakers), Poles, Portugese, Russians, Spaniards, Swedes, Swiss, Syrians, Tristan de Cunhans, Welsh

BLACK AFRICANS

Abyssinians (Amharas), Bantu, Barundi, Batutsi, Bushmen, Congolese, Ewe, Fulani, Gambians, Ghanaians, Hobe, Hottentot, Hututu, Ibo, Iraqi, Kenyans, Kikuyu, Liberians, Luo, Madagascans, Mozambiquans, Msutu, Nigerians, Pygmies, Sengalese, Shona, Somalis, Sudanese, Tanganyikans, Tutsi, Ugandans, U.S. Blacks, "West Africans," Xosa, Zulu

MONGOLOIDS

Ainu, Bhutanese, Bogobos, Bruneians, Buriats, Chinese, Dyaks, Filipinos, Ghashgai, Indonesians, Japanese, Javanese, Kirghiz, Koreans, Lapps, Malayans, Senoy, Siamese, Taiwanese, Tatars, Thais, Turks

AMERINDS

Alacaluf, Aleuts, Apache, Atacameños, "Athabascans," Ayamara, Bororo, Blackfeet, Bloods, "Brazilian Indians," Chippewa, Caingang, Choco, Caushatta, Cuna, Diegueños, Eskimo, Flathead, Huasteco, Huichol, Ica, Kwakiutl, Labradors, Lacandon, Mapuche, Maya, "Mexican Indians," Navaho, Nez Percé, Paez, Pehuenches, Pueblo, Quechua, Seminole, Shoshone, Toba, Utes, "Venezuelan Indians," Zavante, Yanomama

OCEANIANS

Admiralty Islanders, Caroline Islanders, Easter Islanders, Ellice Islanders, Fijians, Gilbertese, Guamians, Hawaiians, Kapingas, Maori, Marshallese, Melanauans, "Melanesians," "Micronesians," New Britons, New Caledonians, New Hebrideans, Palauans, Papuans, "Polynesians," Saipanese, Samoans, Solomon Islanders, Tongians, Trukese, Yapese

AUSTRALIAN ABORIGINES

SOURCE: From R. C. Lewontin in *Evolutionary Biology*, vol. 6. T. Dobzhansky et al. (eds.). Plenum 1972.

* This analysis was made by the population geneticist R. C. Lewontin and was based on data about 17 proteins. Six races and many distinct populations can be recognized.

races (see Table 7-3). The criteria used to distinguish human races include not only features of the body surface, but also less obvious physiological and biochemical differences. In general, the genetic differences between currently recognized human races reflect some degree of reproductive isolation, but there is no denying that because of technologic and sociologic changes human populations are much less isolated from one another than ever before. Although human races do exist, it is questionable whether they have much biologic relevance for our species at the present time. Let us discuss this important point further.

The Importance of Culture in Human Adaptation

Among the more than 1.5 million species of living things that have been described and named, the human species is unique in that its members adapt to the environment primarily by a complicated form of learned behavior called *culture*, which is transmitted from generation to generation by the symbol system of language. For all practical purposes, geographic variation in our species is now irrelevant because people adapt to the environment primarily by means of behavior, whose biological basis is in the brain and is not reflected in superficial differences in body surfaces. When they first evolved, people who had dark-colored skins were better adapted to climatic conditions near the equator than those who had light-colored skins, but few human beings now live under natural conditions. Also, technological advances have assured that at the present time, human beings of various races are as likely to reproduce in one environment as in another. The superficial differences between existing human races are thus largely relics of the past and are not of much functional significance today.

In the evolutionary sense, the significance of races is that under special circumstances some may evolve into new species. The term *subspecies* is used to describe a race that is sufficiently distinct to merit (in the opinion of the person who is classifying) a Latin name in a formal classification. Given enough time and the presence of significant changes in the environment, some isolated groups may become so genetically different from other groups that they can no longer reproduce with one another successfully. When that happens, a new species may evolve.

The amount of variation observed between populations of living human beings probably does not warrant the classification of any of them as official subspecies, especially because racial differences have been blurred as recent years have brought increased mobility and social changes. Nor is there any evidence that groups of people are becoming reproductively isolated from one another and thus may be in the early stages of evolving into a new species. Even the recently abolished castes of India, which were reproductively isolated from one another for at least 3000 years, showed no signs whatsoever that individuals of one caste were incapable of successfully reproducing with members of any other caste or members of any other human population.

But subspecies are officially recognized in extinct members of the human species. Thus, Neanderthalers are officially classified as a distinct subspecies, *Homo sapiens neanderthalensis*. This emphasizes that some Neanderthalers probably interbred with fully modern people, known as *Homo sapiens sapiens*. In fact, modern people may well have evolved directly from isolated groups of Neanderthalers.

In the end, the wondrous diversity of living things, including that of our own species, has probably arisen in large part because of the accumulation of genetic differences between isolated populations. The original source of genetic differences between two individuals is mutation. We now turn to the ways in which heritable changes originate within genetic material and thus produce genetic diversity in any population.

Mutations and Human Diversity

All of the differences between any two living things ultimately originate in mutations-heritable changes that take place within DNA molecules and that are the results of accidents. One of the main ways in which mutations arise is by mistakes that occur during DNA replication. As you know, the sequence of the four bases in one strand of a DNA molecule is complementary to the base sequence of the other strand. Thus, when the two strands separate during DNA replication, each strand provides the base sequence necessary to code for a new complementary partner and the two molecules that result are identical, as long as errors in the copying process do not occur. Most of the time the copying process is completely accurate, but rare mistakes do happen, and these unlikely errors help to furnish the raw material for evolution.

Several types of errors can occur during DNA replication, but in most of them one base pair is substituted for another. This may occur because of the accidental mismatching of two bases that do not usually pair with one another or because of the insertion or excision of one or several base pairs along a particular stretch of DNA. In any case, the end result is that a portion of the genetic code is altered. If the alteration occurs in a structural gene (that is, in a stretch of DNA that codes for a particular protein molecule), then we would expect the protein whose amino acid sequence is coded for by the altered gene to be altered too. Most alterations in protein structure by the kinds of mutations we have been discussing are slight. They usually consist of a single amino acid substitution. For example, it is possible to account for a large part of the naturally occurring variability among the chains of the human hemoglobin molecule by assuming that changes in the amino acid sequence reflect a change in a single DNA base pair in the structural gene in question. Thus, as compared to the corresponding normal chains, each of 34 different alpha chains, 56 beta chains, 4 gamma chains, and 4 delta chains, all of which are known to exist, can be accounted for by single base pair changes.

Mutations that are produced by more extensive changes in DNA molecules are also known to occur. For example, altered DNA molecules can result from unequal crossing over (crossing over usually occurs shortly before cells undergo meiosis, a form of cell division discussed in Chapter 5). The result of unequal crossing over is that hybrid protein molecules can be produced. Thus, persons whose hemoglobin molecules contain a protein chain with an amino acid sequence that begins like that of a normal delta chain and ends like that of a normal beta chain have been reported. These unusual molecules probably result because of unequal crossing over between the structural genes for the human beta and delta chains.

Mutations can also result from physical damage to DNA molecules. Localized regions of DNA molecules can be damaged by energy in the form of radiation of various kinds, including ultraviolet light, x-rays, cosmic rays, and various types of natural radiation from radioactive substances. In order

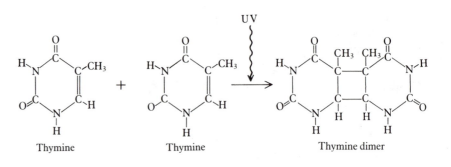

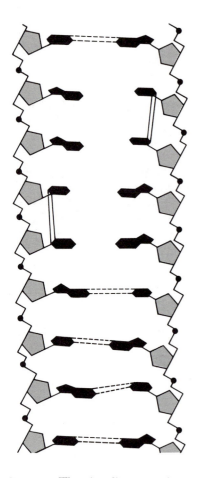

7–12   Thymine dimers can be formed when ultraviolet light interacts with thymine bases in DNA molecules, right. Thymine dimers distort DNA molecules so that they cannot replicate properly, left. (Lefthand portion from "The Repair of DNA," by Philip C. Hanawalt and Robert H. Haynes. Copyright © 1967 by Scientific American, Inc. All rights reserved.)

to compensate for the damaging effects of certain kinds of radiation, organisms have evolved enzymes that repair radiation-damaged segments of DNA. For example, as shown in Figure 7-12, ultraviolet light damages DNA mainly by causing adjacent thymine bases to become tightly bonded to one another, thus distorting the molecule and interfering with its ability to replicate itself. Human cells usually contain an enzyme that has the rather remarkable ability to excise the abnormal regions of DNA, after which other enzymes then repair the excised segment. Rare persons are deficient in the enzyme that excises the abnormal radiation-damaged segments. As expected, their skin is highly sensitive to ultraviolet light and may be severely damaged on exposure to sunlight. (This rare trait is an autosomal recessive disorder known as *xeroderma pigmentosa*).

DNA molecules can also be damaged by heat and by various kinds of chemicals. But before we discuss the relative importance of these and other factors in the production of human mutations, we must mention what happens to mutated genes in the human population once they have arisen.

In terms of natural selection, a person whose DNA contains a mutated gene may be affected positively, negatively, or to no detectable degree. If the influence is negative, the affected person is less likely to reproduce than an unaffected person, or may even die before reaching reproductive age. In either instance, the mutation will be lost from the population unless it recurs spontaneously. If the mutation has no detectable effect on the person, then its persistence in the population is a matter of chance. As we have seen, this is probably true of the great majority of polymorphisms involving protein molecules. Finally, if the mutation has a positive effect it will tend to become more prevalent in a population with the passage of time, mostly because of natural selection, though chance can still play a role. The allele for sickle-cell hemoglobin and its relation to malaria provide a good example.

But almost all mutations are either frankly detrimental or have no detectable effect. This is not surprising. Organisms are already so precisely adapted to survival and reproduction in the environments in which they live that any change is much more likely to be a disadvantage than an advantage. In protein polymorphisms, a high degree of diversity is probably maintained in part by chance and in part by natural selection, though the relative roles of each of these factors in most structural differences of human proteins have yet to be explained.

Estimating the Mutation Rates of Human Genes

Mutations are rare events, and it is particularly difficult to estimate the frequency with which they occur in human populations. There are two main ways, known as the *direct* and *indirect* methods, of estimating the rate at which human genes mutate. Both methods are subject to many sources of error and can yield, at best, only approximate estimates. (Both methods apply only to traits that are obvious and detrimental.)

The direct method of determining the rate at which human genes mutate requires discovering all occurrences of a particular dominant disorder and determining how many of them appear sporadically among the offspring of unaffected parents. As you may recall, about 15 percent of the cases of Marfan's syndrome occur in this way (see Chapter 1), and data from an obstetrical hospital in Copenhagen indicate that mutation may account for a much larger percentage of achondroplastic dwarfism (Figure 7-13).

7–13   Brother and sister achondroplastic dwarfs of the Owitch family as they appeared in 1949 when they arrived in Israel after having spent several years in Auschwitz concentration camp. Their lives were spared because they were used for medical experiments. The autosomal dominant gene (or genes) responsible for this trait does not impair fertility. (United Press International.)

The more obvious sources of error in the direct method are these: First, it is hard to be sure that *all* of the instances of any disorder have been identified, no matter how apparent the disorder may be. Second, it is possible that the trait being studied may result from any of several different mutations, all of which produce the same end result. (This is probably true for achondroplastic dwarfism.) Third, a dominant gene may be incompletely expressed in a particular person because of its interaction with other genes. Thus, a particular trait may suddenly appear in the offspring of parents who, although they appeared unaffected, actually carried the gene for the disorder. Of course, it would then not be a mutation.

Although the direct method applies only to dominant traits, the indirect method of estimating the mutation rate of human genes can be applied to both dominant and recessive traits. The indirect method is based on the assumption that the rate at which mutant genes are added to a population is balanced by the rate at which they are removed by natural selection. This implies that affected persons are less likely to reproduce than those who are unaffected, which tends to reduce the frequency of the mutant gene and to numerically cancel out the number of new mutant genes added by spontaneous mutation during any time interval. The indirect method depends on estimating reproductive fitness, which is a measure of the likelihood that a person affected by a particular trait will reproduce. The indirect method is subject to the same sources of error as the direct method, and the estimation of reproductive fitness thus adds still another source of error, which makes the indirect method of estimating human mutation rates even less reliable than the direct method.

When all of these factors are taken into account, the most reliable estimate is that *a mutation for any particular human gene, or at least for those genes that result in rare, detrimental traits in their mutated forms, is encountered in about one out of every 100,000 to 1,000,000 sex cells.* For the most part, mutations seem to occur about equally often among the sex cells of both sexes. Table 7-4 shows estimates of the mutation rates of certain human genes.

From the total number of autosomal recessive traits that are known and from the rate at which at least some human genes responsible for autosomal recessive traits are known to mutate, it can be calculated that all persons probably carry at least several highly detrimental autosomal recessive genes. This estimate is borne out by the rate at which autosomal recessive traits appear among the offspring of consanguineous matings. Overall, it is estimated that the average person probably carries about five recessive genes that in the homozygous condition could result in death before the person could reproduce.

Although the mutation rates for individual genes are somewhat variable, higher mutation rates for almost all genes can be produced in several ways. For example, increased amounts of radiation, higher temperatures, and the effects of various chemicals all tend to increase the mutation rate, sometimes markedly so. (These factors directly alter DNA molecules, but their effects often show up as damage to entire chromosomes or parts of chromosomes. Phenotypic changes that occur because of altered chromosomes are also considered mutations. See Chapter 1.)

It has been estimated that technologically advanced populations, because of exposure to diagnostic x-rays, radioactive fallout, and natural sources of radiation, have tripled the amount of radiation to which their sex cells are exposed during the reproductive years. But there is evidence that overall, radiation from x-rays, fallout, and other sources does not play a major role in the production of human mutations, at least those with apparent effects.

On the other hand, it is known that for various experimental animals the spontaneous mutation rate increases directly with temperature, as is expected of any process that can be explained in molecular terms. This may have some relevance for human genetics because it has been found that the temperature of the gonads of men wearing pants, especially tight-fitting ones, is higher than that of men wearing kilts or wearing nothing at all.

Various chemicals are known to increase the mutation rate among experimental animals, including fruit flies and mice. Nitrogen mustard is a good example, and because of its biochemical effects on rapidly dividing cells, this toxic substance has been of some benefit in the treatment of certain human cancers. Many other chemicals that many of us encounter in our everyday lives—caffeine, for example—are known to increase the mutation rate in bacteria and in some insects, but evidence that they do so in human beings is not conclusive. Nonetheless, in the end, most human mutations probably result from short-lived, highly reactive chemicals that are produced inside normal living cells (and from errors during DNA replication). These highly reactive chemicals are usually formed as the byproducts of normal metabolism, or, more rarely, may result from the effects of radiation.

In recent years it has been asked whether manufactured chemicals, which are readily available in the environments of industrialized populations, are increasing the rate at which our species' genes are mutating. While environmental chemicals may have had some influence on the mutation rate of

TABLE 7–4   ESTIMATES OF MUTATION RATES OF
CERTAIN HUMAN GENES FROM NORMAL TO ABNORMAL

Trait	Mutant Gene per 100,000 Sex Cells
Autosomal dominants	
Huntington's chorea	<0.1
Nail-patella syndrome	0.2
Epiloia (type of brain tumor)	0.4–0.8
Aniridia (absence of iris)	0.5
Retinoblastoma (tumor of retina)	0.6–1.8
Multiple polyposis of the large intestine	1–3
Achondroplasia (dwarfness)	4–12
Neurofibromatosis (tumors of nervous tissue)	13–25
Marfan's syndrome	0.4–0.6
X-linked recessives	
Hemophilia A	2–4
Hemophilia B	0.5–1
Duchenne-type muscular dystrophy	4–10

some human genes, it is impossible to estimate their overall effect at this time. Mutation rates comparable to those estimated for human beings have also been observed among fruit flies, mice, and other creatures raised under controlled laboratory conditions. It may be true that mutation rates in the human population are not influenced to any presently measurable degree by manufactured chemicals, but rather depend in large part on spontaneous chemical reactions that occur inside normal cells. Nonetheless, it seems wise to continue to question whether or not the benefit of introducing a chemical into a particular environment warrants the risk that the substance will damage the genetic programs of the human and nonhuman populations living there.

Summary

The protein molecules of individuals of the same species are extremely variable. Slightly different forms of almost all human proteins are known to exist, and most of the time slight variations in protein molecules from one person to another do not have ill effects. For most human protein polymorphisms, there is no evidence that natural selection influences the frequencies of the slightly different genes that code for the slightly different proteins. Nonetheless, the gene for the abnormal beta chain of hemoglobin that is characteristic of sickle-cell anemia is maintained at a rather high frequency because natural selection favors the reproduction of heterozygotes in malarious areas.

The worldwide distribution of the ABO blood groups is probably best explained as the result of human migrations. Although various blood groups have been correlated with ulcers, with resistance to certain infectious diseases, and with maternal–fetal incompatibilities, the evidence that natural selection plays a major role in the distribution of the ABO blood groups is not convincing.

The alleles of the MHC code for proteins known as human leukocyte antigens (HLAs), which are located on the surfaces of most kinds of cells. There are four major categories of HLAs, and they exist in many slightly different forms. More than 25 million different combinations of the alleles of the MHC occur in the worldwide human population.

In large populations, random fluctuations in the frequencies of those genes not obviously influenced by natural selection tend to cancel out. But in small populations, significant changes in genetic constitution often occur by chance alone. Thus the Dunkers, who have been reproductively isolated for many generations, have gene frequencies that are different from those of either their West German forebears or their American neighbors.

Most widespread species look slightly different in different geographic areas, and this reflects some degree of genetic difference between populations. Distinct local breeding groups are known as races. In practice most of these local populations blend with one another because many individuals in neighboring populations interbreed.

External features of the human body, such as skin color and body build, are generally more variable from one local human population to another than are less apparent traits, such as slight differences in protein molecules. Racial differences in body surfaces probably became established in ancient, isolated human groups in large part because of natural selection. The number of human races recognized depends on the purposes and opinions of the person who classifies them.

Although human races surely exist, they are generally biologically irrelevant today. This is because most people no longer adapt to the environment primarily by means of body surfaces, but rather by means of language and culture. Human racial differences have been blurred in recent years by increased mobility and social changes.

Mutations are heritable changes that arise within existing DNA molecules, most of them by accident. Mutations that arise during DNA replication usually are produced by the substitution of a single base pair. Unequal crossing over, radiation of various kinds, heat, and numerous chemicals can either produce mutations directly or speed up the rate at which they occur.

Mutations in the human population are particularly hard to detect. Direct and indirect methods of estimating the mutation rate of certain human genes have been devised, but they are subject to many sources of error, and the estimates are very approximate.

Exposure to x-rays, environmental chemicals, natural radiation, and heat surely have some effect on the rate at which human genes mutate, but they may not be the most important factors. Rather, intrinsic errors in the replication of DNA as well as the presence of short-lived by-products of normal metabolism produced inside normal cells probably account for the occurrence of most human mutations.

Suggested Readings

"The Genetics of Human Populations," by L. L. Cavalli-Sforza. *Scientific American*, Sept. 1974. Discusses how the molecular differences within human populations are greater than those between populations.

"Sickle Cells and Evolution," by Anthony C. Allison. *Scientific American*, Aug. 1956, Offprint 1065. How natural selection can maintain an allele that seems to be frankly detrimental.

"The Genetics of the Dunkers," by H. Bently Glass. *Scientific American*, Aug. 1953, Offprint 1062. Describes the evidence for genetic drift among the members of a religious sect.

"Genetic Drift in an Italian Population," by L. L. Cavalli-Sforza. *Scientific American*, Aug. 1969. Discusses the genetic effects of physical isolation and of consanguineous marriages among people who live in Italy's Parma Valley.

"Ionizing Radiation and Evolution," by James F. Crow. *Scientific American*, Sept. 1959, Offprint 55. Discusses the role of x-rays and other kinds of ionizing radiation in the evolutionary process.

Human Diversity, by Richard Lewontin. Scientific American Books, 1982. An eminent overview of the genetic basis of human variability by an expert on population genetics.

Blood Relations: Blood Groups and Anthropology, by A. E. Mourant. Oxford University Press, 1983. Provides authoritative coverage of the worldwide distribution of the ABO blood group.

Chapter 8

The Genetics of Human Behavior

The most variable thing about the human species is the nearly endless variety of ways in which people behave. As was discussed in the preceding chapter, human beings are unique among living things in that they adapt to the environment primarily by a complicated form of learned behavior called culture, which is transmitted from generation to generation mainly by the symbol system of language. Because the biological basis of these complicated learned behaviors lies in the circuitry of the human brain, the capacities for language and culture can be considered genetically determined. However, very little is known about how the behavior of individuals relates to the brain, or how genetic and environmental factors interact throughout a person's life to result in behavior. The problem is that although certain features of human behavior are known to have a genetic basis, it is impossible to assess the relative effects of the environment on the expression of the many genes that influence behavior.

A further difficulty in assessing the genetic basis of many aspects of human behavior is that a person's behavior rarely remains the same for any extended period of time. Most of us do not behave the same way we did 10 years ago (or 20 or 30 years ago), and few of us will be behaving as we do now 10 or 20 years hence. Human behavior is not only complex, hard to define, and difficult to measure, but changeable as well. No wonder the genetic basis of human behavior is poorly understood.

There is very little evidence about the genetic basis of an individual's behavior, and it is usually impossible to accurately assess the environment's influence on it. Nonetheless, the relative contributions of genes and environment are often of considerable interest, largely because of social proposals that bear on human behavior. In this chapter, we shall consider two broad categories of human behavior that have important social considerations: intelligence (or, to be more precise, the genetics of IQ scores) and criminal behavior. As if these two topics were not delicate and heady enough, we shall conclude this chapter and this book by considering how human behavior, natural selection, and other factors may influence human evolution in the future.

The Marked Influence of Single-Gene Defects on Behavior

Let us begin by considering some behavioral abnormalities that result from the presence of a single defective gene. The most striking defective gene that results in abnormal behavior is the one that causes the Lesch-Nyhan syndrome, first described in 1965. As was mentioned in Chapter 6, the abnormal allele for this fatal condition is located on the X chromosome (sex-linked), and its presence results in a deficiency for HPRT. Only males are affected;

This painting by Pablo Picasso ("Girl before a Mirror," 1932, March 14. Oil on canvas, 64 × 51¼) symbolizes the fragmented, emotional consideration that the human species sometimes gives to its own genetic future. (Collection, The Museum of Modern Art, New York. Gift of Mrs. Simon Guggenheim.)

affected females die as embryos. A shocking fact about this fortunately rare disease is that affected boys have abnormal posture and spastic and involuntary movements (among other abnormalities), and they regularly engage in a bizarre form of self-destructive behavior. Unless their teeth are removed, they invariably bite off the tissue of their lips, and they frequently use their teeth to tear at the flesh of their hands, sometimes mutilating themselves severely. The self-mutilation is thought to be involuntary, because they appear to be terrified by their self-destructive activities, and cry out in pain when they bite themselves. Although the exact relationship between the absence of the enzyme HPRT in certain parts of the brain and the presence of this unfortunate behavior remain to be explained, the Lesch-Nyhan syndrome is an example of abnormal behavior that as far as we know is determined by simple genetic factors alone.

Boys who have Lesch-Nyhan syndrome are also usually mentally retarded to some degree. (Mental retardation is defined in a following section.) You will recall that many inborn errors of metabolism, PKU for example, are known to result in some degree of mental retardation. This fact bears witness to the intricacy of the human brain's biochemistry and to its dependence on the interactions of many different genes. In PKU, the abnormal enzyme presumably leads to a biochemical defect in the developing brains of affected persons, and this defect results in mental retardation, though exactly how this occurs is poorly understood. PKU would thus seem to be an example of a behavioral abnormality that is wholly genetic and independent of environmental factors. Yet environmental factors, in the form of special diets, can prevent the mental retardation associated with this condition. Thus, environmental factors sometimes strongly influence a person's behavior even when genetic factors appear to be of overwhelming importance.

Most inborn errors of metabolism are devastating conditions, and their behavioral effects tell us little about the genetic basis of normal behavior. For example, what is the genetic basis of the tendency for compassion, perseverance, altruism, leadership, curiosity, inventiveness, or any of the endless qualities of human behavior that can be singled out, labeled, and measured by a psychological test? It has yet to be determined whether most measurable behavioral differences between human beings result primarily from genetic or environmental factors. There is every reason to expect that such complex and ill defined traits result from the interaction of many genes and many environmental factors. Before we consider the genetics of human behavior any further, let us turn to some methods that can be used to assess the relative roles of genes and environment in the determination of behavior.

Nature and Nuture—The Concept of Heritability

As discussed in preceding chapters, the inheritance of those traits that show continuous and gradual variation within a population usually depends, not on a single pair of alleles, but on many pairs that in each person interact both with the rest of the individual's genes and with the environment. Consider, for example, the multitude of genetic and environmental factors that interact to determine how tall a person is. Height is influenced by many genes, such as those that code for growth hormone, for intestinal enzymes that digest food and thus provide the body with building blocks for growth, and for the rate at which calcium is deposited in the long bones of the legs. But environmental factors also make an important contribution to a person's stature. For example, the absence of sunlight can result in inadequate synthesis of vitamin D, which may result in the slowing down of proper bone growth. Chronic poor nutrition in childhood can also influence adult height. The end result of all of these factors affecting height, and of the many others that must also play a part, is that in the worldwide human population, and in any random sample of it we wish to single out, height is normally distributed. That is, most individuals in any human population are not far away from being of some average height, and although there are very tall and very short persons, there are fewer of them than persons of average stature (Figure 8-1).

Plant and animal breeders have known for centuries that continuously varying traits are influenced by the environment to different degrees. Because the breeders are concerned with establishing true-breeding lines of plants and animals that have desirable characteristics, it is important for them to be able to assess the relative effects of genes and environment on characteristics they consider desirable, such as copious milk production from cows, large eggs from hens, and long coats on woolly sheep. All of these characteristics are influenced by many genes and many environmental factors, and breeders have found the statistical concept of *heritability* useful in assessing the relative role of the genetic factors that they would like to "breed into," and thereby genetically improve, the breeds that they have already developed. *Heritability* is that proportion of the phenotypic variation in any population that can be attributed to genetic factors.

How does one assess the relative effects of genes and environment on a continuously varying trait that is desirable? In practice it is very difficult to estimate heritability, but there are several ways of going about it. For example, heritability can be estimated by mating individuals from a given position in the normal distribution curve for a particular trait, and then examining the distribution of the trait in their offspring, who must be raised under strictly controlled environmental conditions. If all of the variation in a particular trait in a particular population is due to genetic factors alone, the heritability is assigned the number 1 (100 percent), and the average value of the trait in the offspring is equal to the value of the position on the curve from which the parents were selected. On the other hand, if all of the variation is due to environmental factors, the heritability is 0 (0 percent), and the average value of the trait in the offspring is the same as the average value in the population from which the parents were selected. Most continuously varying traits have heritabilities between 0 and 1 that can be estimated by determining the difference between the average values in selected parents and their offspring (Figure 8-2).

Heritability can also be estimated by experiments in inbreeding, over at least several generations and under controlled environmental conditions, of animals that are related. This method of estimating heritability depends on the fact that different kinds of relatives (brothers and sisters, cousins, etc.)

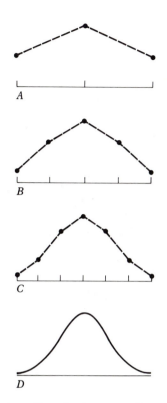

8–1   How the frequency distributions of phenotypes are related to the number of pairs of alleles involved. A, one pair of alleles distributed over three phenotypes. B, two pairs of alleles distributed over five phenotypes. C, three pairs of alleles distributed over seven phenotypes. D, an infinite number of pairs of alleles distributed over a continuous array of phenotypes. (This curve is the *normal distribution curve* that characterizes most biological populations. Compare with Figure 1-17.) (From Curt Stern, *Principles of Human Genetics*, 3d ed. W. H. Freeman and Company. Copyright © 1973.)

8–2   Heritability can be estimated by breeding experiments that
allow one to calculate the ratio of
the gain to the selection differential.
In the three top curves, the difference between the mean (average) of
the selected parents and the mean of
the population they were selected
from (the *selection differential*) is the
same. In each case, the offspring
that are produced have characteristics that form a normal distribution
curve, and the change in mean between parents and offspring (the
gain) can be used to estimate the
heritability of a particular trait,
which is represented by h^2. (From
I. Michael Lerner and William J.
Libby, *Heredity, Evolution, and Society*, 2d Ed. W. H. Freeman and
Company. Copyright © 1976.)

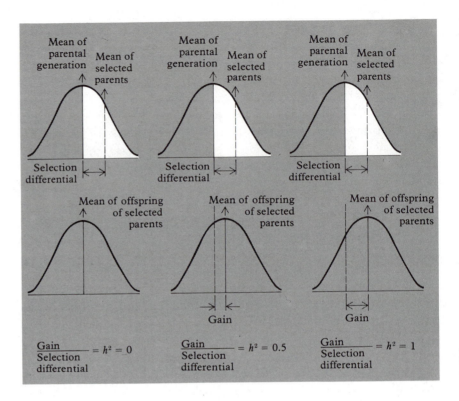

share to different degrees the genes that influence the trait under consideration. Figure 8-3 shows the range of heritabilities, as determined largely by
experiments in inbreeding, for various economically important traits of the
domestic chicken.

But heritability is a statistical concept that applies to populations, not to
individuals. For example, as shown in Figure 8-3, the average heritability
for the weight of hens' eggs is about 0.75 (75 percent). A heritability of 0.75
does not mean that for each hen egg weight is determined three-fourths by
heredity and one-fourth by environment. What it does mean is that overall
three-fourths of the total variation in the weight of hens' eggs is associated
with genetic differences.

Although heritability is never easy to measure, it is particularly difficult
to estimate in human populations. There are several reasons for this, the most
obvious of which is that people are not experimental animals that can be
selectively mated, highly inbred, or raised under strictly controlled conditions. A less obvious reason why heritability is difficult (in fact, impossible)
to determine accurately in human populations is that it depends not only on
genes and environment but on the *interaction* between the two. In human
populations, information concerning the exact contribution that the interaction of genes and environment makes to the heritability of human traits is
often meager, if not nonexistent.

Nonetheless, the heritability of certain human traits that depend on many
genes can be crudely estimated from the rates at which the traits occur among
close relatives of a given person, compared with the population at large. Thus,

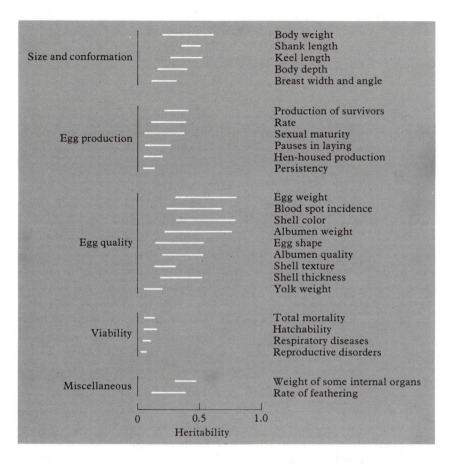

8–3   The range of heritabilities reported for various economically important characteristics of the domestic chicken. (From I. Michael Lerner and William J. Libby, *Heredity, Evolution and Society*, 2d Ed. W. H. Freeman and Company. Copyright © 1976.)

it is estimated that the heritability of height among white women in the United States is about 0.94 (94 percent), while that of hip circumference is 0.66 (66 percent), and that of weight is 0.42 (42 percent).

Heritability estimates apply only to a given population at a given time and are subject to change. Some biologists and statisticians have suggested that the concept of heritability, because of its built-in limitations, should not be applied to human populations at all. There is probably some virtue to this argument, especially for estimates of poorly defined or poorly understood traits such as intelligence (see the following discussion). But fortunately, nature has provided us with at least one fairly accurate way of estimating the effects of genes and environment in human populations, whether one has much faith in the concept of heritability or not. Human beings sometimes are produced in almost duplicate copies known as identical twins, who for all practical purposes are genetically identical.

Nature and Nuture—Twin Studies

There are two kinds of human twins: one-egg, or *monozygotic*, twins and two-egg, or *dizygotic*, twins. Dizygotic twins are produced when two eggs, rather than the usual one, are released at ovulation and both are subsequently

TABLE 8–1

THE PERCENTAGE OF CONCORDANCE AMONG TWINS FOR SOME TRAITS THAT DEPEND ON MANY GENES AND MANY ENVIRONMENTAL FACTORS

Observed Disease or Behavior	Percentage Concordance	
	MZ Twins	DZ Twins
Tuberculosis	54	16
Cancer at the same site	7	3
Clubfoot	32	3
Measles	95	87
Scarlet fever	64	47
Rickets	88	22
Arterial hypertension	25	7
Manic-depressive syndrome	67	5
Death from infection	8	9
Rheumatoid arthritis	34	7
Schizophrenia (1930s)	68	11
Criminality (1930s)	72	34
Feeble-mind-edness (1930s)	94	50

SOURCE: From *Heredity Evolution and Society,* 2d ed., by I. Michael Lerner and William J. Libby. W. H. Freeman and Co. Copyright © 1976.

fertilized by different sperm. The genetic relationship between dizygotic twins is thus the same as that between brothers and sisters who are not twins. On the other hand, monozygotic twins originate from a single fertilized egg that, after having been fertilized by a single sperm, splits into two at a very early stage of development and thus results in two individuals who are, barring somatic cell mutations during development and in later life, genetically identical (Figure 8-4). (Alterations in DNA sequences that occur during the synthesis of antibodies also introduce slight differences into the genetic programs of monozygotic twins.)

Statistical studies of the differences between identical twins sometimes focus on complex characteristics that depend on many genes and many environmental factors but that nonetheless do not show continuous and gradual variation. The relative role of genetic factors in the expression of such all-or-none traits can be estimated by the degree to which the twins are concordant or discordant. When both twins have a particular trait, they are concordant, and when only one does, they are discordant. Table 8-1 shows the degree to which monozygotic and dizygotic twins are concordant for various abnormal conditions that depend on the interaction of many genes and many environmental factors. The percentage of concordance provides an estimate of the degree to which a particular condition is genetically influenced.

A good example of an abnormality of human behavior that probably depends on the interaction of many genes and a multitude of environmental factors is *schizophrenia.* Contrary to popular usage of the word, this serious behavioral abnormality, of which several types are officially recognized, is not manifested by "split personality," but rather by a split between thoughts and feelings and by a loss of contact with the environment (there are many other features). Schizophrenia is by no means rare. It is estimated that at least 2 million persons in the United States alone either have schizophrenia or will develop symptoms of the disease sometime during their lives.

Although the issue is far form settled, most geneticists are convinced that genetic factors influence the development of certain types of schizophrenia.

TABLE 8–2   CONCORDANCE OF SCHIZOPHRENIA IN TWINS

Country	Year	Monozygotic Twins		Dizygotic Twins	
		Number of Pairs Studied	Percent Concordance	Number of Pairs Studied	Percent Concordance
Denmark	1965	7	29	31	6
Germany	1928	19	58	13	0
Great Britain	1953	26	65	35	11
Japan	1961	55	60	11	18
Norway	1964	8	25	12	17
United States	1946	174	69	296	11

SOURCE: From *Heredity, Evolution and Society,* 2d ed., by I. Michael Lerner and William J. Libby. W. H. Freeman and Co. Copyright © 1976.

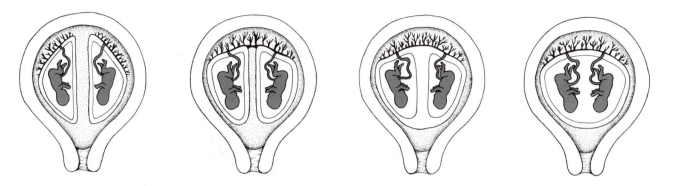

Part of the evidence for this comes from studies of twins. As shown in Table 8-2, both monozygotic twins develop schizophrenia much more commonly than do both dizygotic twins. While such studies clearly indicate a role for genetic factors in the development of schizophrenia, they are not conclusive. The reason is that each pair of twins studied was raised in the same household by the same parents, and environmental factors (such as inconsistencies in the behavior of the parents) are known to be a major factor in the development of at least some kinds of schizophrenia.

Identical twins have nearly all of their genes in common, so any variation between two identical twins is in large part due to the effects of the environment. Most of the time identical twins are raised in the same environment, and under such circumstances they tend to resemble each other strongly both physically and in at least some aspects of behavior (Figure 8-5). But sometimes monozygotic twins are raised apart from one another in different environments, and when this happens a study of the differences between the two individuals can provide a measure of the genetic and environmental components of certain human characteristics.

Studies of the physical and behavioral characteristics of identical twins raised apart sometimes reveal mind-boggling similarities. Consider a few that came to light during a recent study at the University of Minnesota. A pair of 39-year-old female twins who were separated shortly after birth each arrived at the study center wearing seven rings (on the same fingers); each also wore two bracelets on one wrist, and a watch and a bracelet on the other. One twin had a son named Richard Andrew; the other had a son named Andrew Richard. A pair of 39-year-old male twins, who were separately adopted as 4-week-old infants and had never met, had the following features in common: Both were named Jim, both had law enforcement training and worked part-time as deputy sheriffs, both had dogs named Toy, both had married and divorced women named Linda and had remarried women named Betty, both drove Chevrolets, both liked mechanical drawing and carpentry, both had very similar smoking and drinking patterns, and both bit their fingernails to the quick. One twin had a son named James Allan and the other's son was named James Alan. The significance (if any) of these astounding similarities remains to be determined. More importantly, each pair of twins in the study was subjected to six days of physical and psychological testing; in general, their test scores are as close as would be expected for the

8-4   Monozygotic and dizygotic twins may share the prenatal environment to different degrees because of the arrangement of the membranes surrounding the fetuses.

8–5   Top, identical twins Bruno and Giorgio Schreiber at about two years of age. Even the adult twins cannot tell who is who in this photo. Center, as adults, the twins continued to look very much alike. Bottom, a recent photo. Bruno (left) and Giorgio (right) are now professors of zoology at the Universities of Parma (Italy) and Belo Horizonte (Brazil), respectively. (Courtesy of B. Schreiber.)

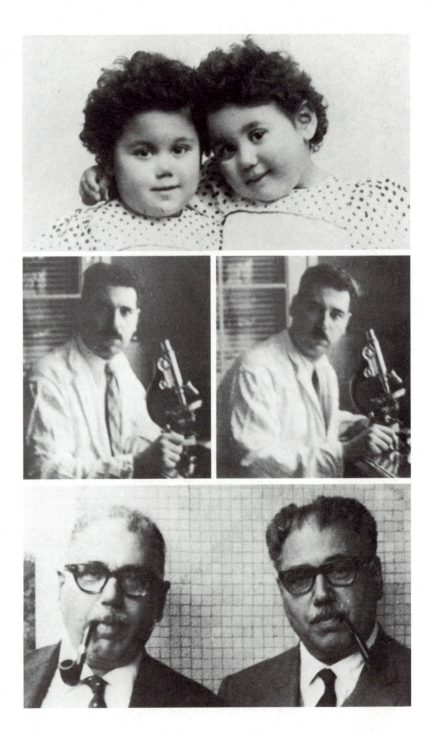

same person taking the same test twice. (Of all of the tests administered to the twins, the highest concordances were in their scores on **IQ** tests.)

Although studies of monozygotic twins raised apart seem to provide an ideal way of demonstrating the degree to which a given trait or behavior is genetically determined, they have at least two serious flaws. First, identical twins raised apart are rare, and the number of pairs studied is therefore very small. It could be true that the **IQ** (intelligence quotient) scores of two randomly selected individuals have the same chance of being similar as those of identical twins raised apart. In other words, the striking similarities between the twins may be no more common than coincidences that occur between two persons selected at random from a very large group. Second, most of the time, the environments in which the separated twins grow up are not very different. In one large study of 44 pairs of twins separated soon after birth, all but 4 of the pairs were raised by close relatives, friends, or neighbors of the mother or father. The similarity of the environments in which the twins are raised makes the interpretation of observed similarities between the twins, especially behavioral ones, very difficult.

The Genetics of Mental Retardation

Twin studies involving individuals of subnormal mental development, and for whom no environmental cause such as brain damage during birth can be found, have consistently yielded very high concordances. Mental retardation or deficiency is hard to define. For our purposes, a *mentally retarded individual* is defined as one who, because of subnormal mental development, is not capable of an independent existence and whose IQ is 69 or lower. (See the next section for further discussion of IQ.)

Mental retardation is not rare. In the United States, about two or three per 100 individuals are mentally retarded. As shown in Figure 8-6, about 20 percent of the cases of mental retardation can be attributed to environmental factors such as infections of the brain, prolonged lack of oxygen during birth, and so on. About 22 percent of the cases are caused by known dominant or recessive alleles or by various combinations of known detrimental alleles. This category includes many inborn errors of metabolism, and affected in-

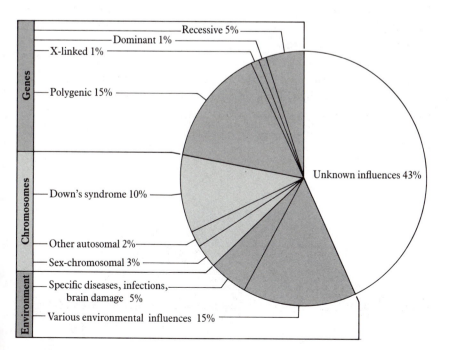

8–6   Estimates of the relative incidence of types of mental retardation that result from genetic and environmental factors. (After Penrose, in Wendt, ed., *Genetik und Gesellschaft*, Wiss. Verlagsges., Stuttgart, 1971.)

dividuals tend to be profoundly retarded. Chromosomal abnormalities account for about 15 percent of the cases of mental retardation. The most frequently encountered abnormality is Down's syndrome, which accounts for about 10 percent of all mentally retarded individuals in the United States. Individuals with Down's syndrome can range from "profoundly retarded" (IQ less than 20) to "mildly retarded" (IQ from 52 and 69), and those individuals with the highest IQs are usually genetic mosaics for Down's syndrome. About 43 percent of the total number of cases of mental retardation are of unknown cause and the recurrence risks have not been reliably determined, which makes accurate genetic counseling very difficult.

The Genetics of IQ Scores

IQ stands for *intelligence quotient*. The concept of IQ was introduced in 1903 by Alfred Binet as a means of allowing teachers in Paris to allocate children to their correct grade in school and to identify those individuals whose mental development seemed to be lagging behind and who therefore might benefit from special attention. Retarded individuals were easily identified by comparing their mental age (as judged by their test scores and by their teachers) with their chronological age. For example, if a 12-year-old child's mental development was judged to be the same as that of an average 9-year-old, then the child's intelligence quotient was 9/12, which corresponds to an IQ of 75. The choice of an IQ of 69 as the cutoff between normal and retarded is, of course, arbitrary.

Not long after Binet's tests were introduced, the concept of IQ took on much broader scope. In the United States, the test questions were expanded and modified to produce the *Stanford-Binet IQ test*, which is still the standard IQ test against which all newer tests have been validated. The original Stanford-Binet test stressed memory, the ability to recognize similarities and differences among objects, and vocabulary. The tests were designed so children who were judged by their teachers as "smart" did better than those who were judged as "average" or "dull." Test scores were adjusted for age by dividing the actual score by a correction factor, thus standardizing the test for each age group. More recent tests have also been standardized for sex; that is, questions on which either boys or girls consistently score differently have been eliminated. It is also claimed that some tests have removed cultural,

8–7   Distribution of IQ scores in the white American population. The test has been standardized so that the mean score is 100. The percentage of the population falling within each range of scores is given above the horizontal axis. For example, 7 percent of the persons have an IQ between 70 and 80 and between 120 and 130. (From Richard Lewontin, *Human Diversity*. Scientific American Books. Copyright © 1982.)

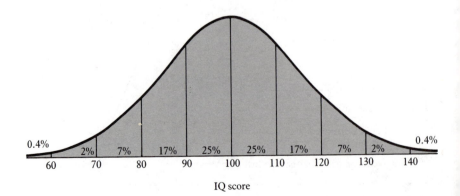

Genetic and nongenetic relationships studied		0.00	0.10	0.20	0.30	0.40	0.50	0.60	0.70	0.80	0.90	Expected values
Unrelated persons	Reared apart											0.00
	Reared together											
Foster-parent-child												0.00
Parent-child												0.50
Siblings	Reared apart											0.50
	Reared together											
Twins — Two-egg	Opposite sex											0.50
	Like sex											
Twins — One-egg	Reared apart											1.00
	Reared together											

8–8   A summary of correlation coefficients concerning IQ scores compiled by L. Erlenmeyer-Kimling and L. F. Jarvik from various sources. The horizontal lines show the range of correlation coefficients for intelligence between persons who are related to various degrees either by genes or by environment. (After I. Michael Lerner and William J. Libby, *Heredity, Evolution, and Society*, 2d Ed. W. H. Freeman and Company. Copyright © 1976.)

racial, and linguistic bias, but not everyone agrees that this has been accomplished, or that it is even possible to do so. The distribution of IQ test scores in the American white population is shown in Figure 8-7.

Intelligence quotient is best defined not as a measure of a person's "intelligence," but as a measure of a person's ability to perform well on IQ tests. This definition is appropriate because it is impossible to define intelligence to everyone's satisfaction, and even if we could define the term, we could not measure it by the same yardstick in all human populations. (To get some idea of the difficulties, how would you measure the intelligence of European college students as compared with that of persons of the same age group who live in the Amazonian rain forest, where they survive without the benefits of the technological achievements of the culture of the former group?) IQ tests were developed to measure some of the mental aptitudes of white middle-class people who live in North America and Europe. IQ test results are informative in that they provide a reliable way of predicting success in school. (In fact, "school-performance predictor tests" is a much more accurate description than "intelligence quotient tests.") Thus, the ability to score high on IQ tests is important only insofar as it is important to achieve success in school within a given cultural framework. Judgments of what constitutes success are always open to question and subject to change. Whatever its ultimate importance, within white middle-class populations 'the ability to score high on IQ tests appears to be appreciably influenced by genetic factors.

The main problem in trying to figure out the genetic basis of IQ scores is that it is impossible to sort out accurately the genetic and environmental influences that interact to determine how a person performs on IQ tests. Nonetheless, there are several ways of roughly estimating the genetic component of IQ scores. Monozygotic twins raised apart are one source of information, and studies of them indicate a high degree of heritability. For traits that show continuous and gradual variation, whether between twins or among other relatives, the degree of similarity is best measured by the value of the statistical quantity known as the *correlation coefficient*, whose numerical value can vary between 0 and 1. In general, a high correlation coefficient suggests a high heritability. See Figure 8-8.

As was mentioned in the discussion of the recent University of Minnesota study, the IQ scores of monozygotic twins raised apart are highly correlated.

Nonetheless, one pair of male twins in this study had one of the largest differences in IQ scores ever reported for identical twins (24 points), and as mentioned earlier, the validity of studies assessing the relative contributions of genes and environment to differences and similarities among twins is open to question. That environmental effects can influence the IQ of twins is evident in that the average correlation of the IQ scores of identical twins raised together is .88, whereas that of identical twins raised apart is .75.

The genetic component of IQ scores can also be estimated by comparing the test scores of adopted children with those of the biological and foster parents. Table 8-3 lists the correlation coefficients for IQ scores between mothers and their children in three large adoption studies. Because the test scores of biological mothers and their children are more highly correlated than those of foster mothers and their adopted children, these studies seem to provide good evidence for genetic determination of IQ test scores. But there are some caveats. First, in adoption studies it is essential that children of parents whose IQs are high are distributed at random among foster mothers and fathers, but in fact such children are usually adopted by upper-middle-class families, whose members tend to have higher than average IQs (see below). Second, the very act of adoption significantly raises the IQs of practically all adopted children. In a British study, children were found to have an average IQ of 105 if they were placed in an orphanage as infants and stayed there until they were five years old, while adopted children with similar backgrounds had an average IQ of 115.

We have already discussed the role of nturition in the development of normal mental capacities, and many other environmental factors are known to affect not only normal mental development but IQ scores as well. Among the most important environmental components are psychological and social factors, and several are known to influence IQ scores. First, as shown in Figure 8-9, there is evidence that offspring of larger families tend, overall, to have lower IQ scores than those of smaller families. This had been interpreted (though not without criticism) as evidence that large families may provide an environment that is less satisfactory to the development of higher IQ scores. Second, ongoing studies of the effects of the environment within a given social class indicate that the presence of an "enriched," as opposed

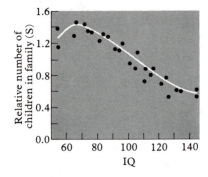

8–9   Data concerning the relation of family size and average IQ. The relative number of children in each family is the actual number of children divided by the average number of children per family in the whole sample. (From Kenneth Mather, *Human Diversity*. Free Press 1964.)

TABLE 8–3   OBSERVED CORRELATIONS IN IQ AND EDUCATIONAL ATTAINMENT BETWEEN PARENTS AND CHILDREN IN THREE STUDIES OF ADOPTION

Study	Correlations		
	Foster Child with Foster Mother	Biological Child with Biological Mother	Foster Child with Biological Mother
Burks	.19	.46	—
Leahy	.20	.51	—
Skodak and Skeels	.02	—	.32

SOURCE: From *Human Diversity,* © by Richard Lewontin. Scientific American Book, Copyright © 1982.

TABLE 8–4 DATA ABOUT THE RELATION BETWEEN OCCUPATIONAL STATUS OF PARENTS AND THE AVERAGE IQ OF THEIR CHILDREN

Occupational Status of Parents	Average IQ of Children		
	United States at Large	USSR	Chicago
Professional	116	117	119
Semiprofessional	112	109	118
Clerical and retail business	107	105 ⎱	
Skilled	105	101 ⎰	107
Semiskilled	98	97	101
Unskilled	96	92	102

SOURCE: From *Heredity Evolution and Society,* 2d ed., by I. Michael Lerner and William J. Libby. W. H. Freeman and Co. Copyright © 1976.

to a "deprived," environment favors the development of higher IQ scores. This reflects the fact that most children who have lower IQ scores come from homes in which the kinds of mental activities measured by IQ scores are not emphasized. And third, the children of persons who have higher occupational status tend to have higher IQs than children whose parents are of lower occupational status. As shown in Table 8-4, this is true of populations studied in the United States and Russia.

The fact that IQ scores and occupational status are strongly correlated does not mean that the ability to score high on IQ tests *causes* a person to be of high occupational status. The lightly shaded bars in Figure 8-10 show the

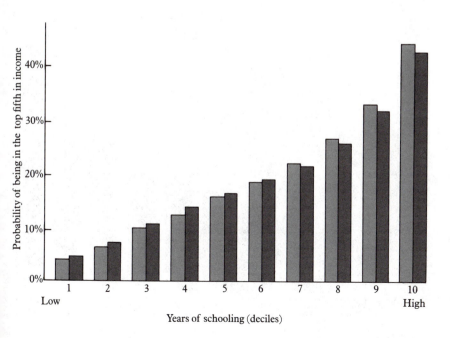

8–10   Relation between the probability of high income and years of schooling for the population as a whole (lightly shaded bars) and for only persons having average IQs of about 100 (darkly shaded bars). (From Richard Lewontin, *Human Diversity.* Scientific American Books. Copyright © 1982.)

probabilities that persons from the U.S. population at large will be in the top 20 percent in income given various amounts of schooling. Because high IQ scores correlate with success in school and with high occupational status and income, does it not follow that high IQ test scores cause economic success, as measured by occupational status and income? No, it does not. The darkly shaded bars in Figure 8-10 show the probabilities of being in the top 20 percent in income when only persons of average IQ are considered, which for the U.S. population is near 100. As you can see, there is very little difference. A man of average IQ was 10 times more likely to have a high income if he was in the top 10 percent of schooling than if he was in the bottom 10 percent. Years of schooling and economic success are thus highly correlated regardless of IQ scores. The probability of economic success and coming from a high socioeconomic class are also highly correlated, as shown in Figure 8-11. Men whose fathers are in the highest (top 10 percent) social stratum in the United States are 10 times more likely to have high incomes than those from the lowest social stratum. Once again, the probabilities change very little when we consider only men of average IQ. The data in these two graphs make it clear that if IQ tests are considered to measure some intrinsic "general intelligence" that underlies success, as reflected in occupational status and income, then the tests are a failure.

Although the reasons for the variation observed between the IQ scores of whites of different socioeconomic classes are poorly understood and sometimes hotly disputed, the controversy surrounding them is dwarfed by the ignorance and confusion surrounding the average IQ scores of various human races, particularly those of blacks and whites in the United States. The difference between the average IQ scores of blacks and whites in the U.S. population is about 15 IQ points. The question is: What does this difference mean, and how important is it?

8–11   Relation between the probability of high income and high socioeconomic background for the population as a whole (lightly shaded bars) and for only persons having average IQs of about 100 (darkly shaded bars). (From Richard Lewontin, *Human Diversity*. Scientific American Books. Copyright © 1982.)

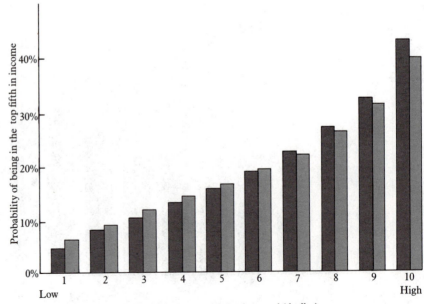

Family socioeconomic background (deciles)

TABLE 8–5   INDEXES OF ECONOMIC AND SOCIAL STATUS OF
AMERICAN BLACKS AND WHITES

	1950		1970	
	Whites	Blacks	Whites	Blacks
Median family income	$3,445	$1,869	$10,236	$6,516
Percent completed high school	33.6	12.1	57.2	35.4
Percent managers and technicians	17.0	3.8	22.5	12.6
Percent unemployed	4.9	9.0	4.5	8.2
Infant and fetal mortality (per 1000 births)	63.3	104.5	44.0	76.9
Life expectancy of males (years)	66.5	59.1	68.0	61.3

SOURCE: From the U.S. Bureau of the Census.

Some white academicians, among whom Arthur Jensen and William Shockley are perhaps the most vocal, have asserted that most, if not all, of the differences in IQ scores between black and white children can be attributed to genetic factors. But as we have seen, it is impossible to sort out the exact contributions of genetic and environmental factors to IQ scores *within* either the black or the white American population, so it is clearly impossible to make meaningful comparisons *between* the two. No matter how high the heritability of IQ scores actually is, the fact remains that, given the obvious environmental differences between populations of middle-class whites and ghetto-dwelling blacks, we would *expect* to find a difference in their performances on IQ tests. In the face of vast environmental differences, it seems premature at best to attribute the expected IQ score difference entirely to systematic genetic differences between the two populations.

In the opinion of many geneticists, the black–white differences in IQ scores most likely reflect, not genetic, but rather cultural and other environmental factors, such as nutrition, family size, and the psychological impact of racism. Table 8-5 lists some important differences in economic factors, social status, and vital statistics between black and white Americans. Studies comparing black and white youngsters who were raised in similar environments are rare. One such study involved children in a British orphanage whose parents were classified as black, white, or "mixed." The children were placed in residential nurseries, usually before 1 year of age, and, after they had been there for at least six months, they were given three nonverbal IQ tests. As shown in Table 8-6, there are no significant differences between the three groups of children. Most of the differences in IQ scores between black and white Americans could probably be eliminated by making the environments the same for both groups.

In the end, IQ scores measure an aspect of human behavior that, like most other features of the ways in which people behave, is desirable in the opinion of some people and undesirable in the opinion of others.

We now turn to an aspect of human behavior that is clearly undesirable to most people, and that has been said by some to be associated with the

TABLE 8–6   SCORES OF
RESIDENTIAL NURSERY
CHILDREN ON THREE
NONVERBAL IQ TESTS

Children of:	Test 1	Test 2	Test 3
White parents	103	98	101
Black parents	107	98	106
Mixed parents	106	99	110

SOURCE: From B. Tizard, *Nature*, vol. 247, 1974, p. 316.

presence of an extra Y chromosome. We will now discuss the relationship between criminal behavior and the chromosome constitution XYY.

The Behavior of Men Whose Sex Chromosomes Are XYY

We have already discussed the way in which males whose sex chromosomes are XYY may be produced from various accidents that occur during cell division. XYY males are born with surprising frequency. It is estimated that at least one in about every 1000 newborn males has XYY sex chromosomes. Most XYY men are taller than those whose chromosomes are XY and some of them have severe acne, but most of them appear otherwise normal.

The controversy that has surrounded the XYY genotype concerns criminal behavior. In 1965 it was reported that the genotype XYY was encountered among men in a certain wing of a mental-penal institution (namely, the Carstairs maximum security hospital in Scotland) at a much higher rate than among the general population. (Seven out of 197 men were XYY, which is about 36 times their frequency in the population at large.) Since that time, other well-documented studies have been published in other countries. The conclusion, though hotly contested by some, seems inescapable to others. It is this: Men of sex chromosome constitution XYY are somewhat more likely to be incarcerated in a mental-penal institution than men whose sex chromosomes are XY. (In general, the offenses of XYY men are similar to those of XYs, and contrary to some earlier reports, XYY men do not appear to be concentrated among the most dangerous, aggressive, or violent inmates.)

Nonetheless, the great majority (at least 98 percent) of men in mental-penal institutions have XY sex chromosomes, and only a small proportion of the total number of men who have sex chromosome constitution XYY engage in criminal behavior. Based on the percentage of XYY men in mental or penal institutions, compared with their incidence in the general population, it is estimated that at least 96 percent do not behave in ways that result in their being institutionalized.

Criminal behavior is so widespread and affects the lives of so many non-criminals that the relation between a slightly increased tendency for criminal behavior and the presence of an extra Y chromosome may be of more than academic interest. But at the same time, the issue is surrounded by prejudice and confusion. As usual, the problem is that it is impossible to accurately assess the effects of genetic and environmental factors. Environmental factors, such as growing up in a ghetto or associating with criminals, are of great importance in the development of criminal behavior.

Now consider some of the moral and ethical difficulties that would arise in a genetic screening program designed to detect XYY males in the general population. Suppose an XYY male infant has just been identified by prenatal diagnosis. What would you advise the parents to do? What would you expect the parents to do because of your advice? Would you tell the child that he is an XYY and thereby implant in his mind the notion that he may grow up to be criminal? Would you mind if the chromosomes of your own male offspring were routinely screened for the presence of an extra Y chromosome without your knowing about it?

The only XYY screening program in the United States has, at least for the present, been shut down because of unrelenting pressure from persons opposed to XYY screening. In the spring of 1975 the faculty of Harvard Medical School, brought to caucus by geneticists and psychiatrists interested in continuing a screening program that had been in existence since 1968, voted 200 to 30 to continue the project. Nonetheless, a few months later the screening project was shut down because those in charge of it said that they were worn out by the pressures of some activist groups that opposed XYY screening.

How should we proceed in this delicate area that may or may not have some genetic basis and that may or may not be subject to potential eugenic measures? Clearly, that depends on the judgments of individuals, and opinions on this matter are abundant and often strongly felt. Perhaps it would be appropriate to follow the example of molecular geneticists who, when recently confronted by the potentially disastrous effects of their research involving recombinant DNA, convened a series of international conferences to discuss the problem. Any such conference on XYY chromosomes should probably include not only geneticists, sociologists, criminologists, social philosophers, and other humanists, but also persons of sex chromosome constitution XYY and representatives of well-informed groups that have strong interest in or opposition to XYY screening.

As if all of the uncertainty surrounding the genetics of IQ scores and the behavior of XYY males were not enough, we now turn to a brief consideration of how natural selection and other factors may affect human populations in the future. (In the discussion of future human evolution that follows, you should remember that, although speculation is sometimes dangerous, sometimes rewarding, and always fun, in the end it is just speculation.)

Human Evolution in the Future

Like all other living things, human beings have evolved and are evolving. But understanding how evolution works does not allow us to make predictions about the future course of the evolution of our own species, or of any other. Nonetheless, we can do some cautious, perhaps meaningful, speculating about the future evolution of the human species, provided that we base our speculations on the assumption that natural selection is at work in the human population today and will be in the future.

Natural selection is at the core of the theory of evolution; in essence it consists of differential reproduction within populations in which individuals differ from one another genetically. Because of natural selection, individuals best suited to survive in a given environment leave more descendants than those who are not so well suited, and because of the effects of natural selection, living things are precisely adapted to their environment. It has been argued that natural selection is no longer important in the evolution of the human species because few human beings now live in "a state of nature," where natural selection can result in the maintenance of human adaptations of benefit in local environments. Moreover, as you will recall from our discussion of human races, differences in body surfaces (which are obviously influenced by natural selection) are now largely irrelevant to our species because people adapt to the environment primarily by means of behavior, not by means of their body surfaces. But this does not mean that natural selection is no longer a major factor in human evolution. In fact, there are at least three ways in which natural selection operates in human populations today—prenatal selection, postnatal selection, and fecundity selection. Let us now discuss briefly each of these ways in which natural selection is at work in human populations today and speculate on how they may affect the future evolution of our species.

Prenatal selection refers to any genetic and environmental factors that result in death sometime between fertilization and birth. As you know, prenatal death is frequently the result of genetic factors, such as abnormal chromosome constitutions or homozygosity for disadvantageous allelles; as such, it is mostly beyond our present reach. Although it is true that genetic counseling and prenatal diagnosis have some influence on prenatal selection in some human populations, any overall effect on the worldwide population has so far has been, and will probably continue to be, meager.

Postnatal selection occurs when infants born alive fail to survive to reproductive age because of some genetically determined defect. Included in this category are many of the serious inborn errors of metabolism and other genetic diseases that were previously discussed. Although the effects of modern medical treatment have been spectacular and of enormous benefit to some persons and their families, most geneticists agree that medical treatment has so far had little effect on the average gene frequencies in the global human population. Nonetheless, it must be admitted that the human species could be affected in a very adverse way by the "genetic load" of detrimental alleles that has already begun to build up because of human intervention and that will surely continue to do so in the future.

Fecundity selection, which may take place both within and between populations, occurs when some genetically distinct members of the population leave relatively more descendants than others. Fecundity selection between human populations at the present time favors people who live in the less highly developed regions of the world, especially in Latin America, Africa, and Asia, but this pattern may change before too long as effective means of birth control become more widely available. Fecundity selection within human populations also occurs, but is hard to measure and subject to frequent changes. Overall, fecundity selection, although it does occur both within and between human populations, provides us with little basis from which to predict the future course of human evolution.

In the end, the overall effect of natural selection within the human population today thus appears to be maintenance of the human species as it is at present. That is not surprising, because natural selection is usually a stabilizing influence that tends to put a brake on rapid or extreme evolutionary changes. This is because the extreme phenotypic variants within any population are much less likely to leave descendants than other members of the population.

Certain evidence in the fossil record also suggests a stabilizing influence of natural selection. The human species in its present form has been in existence for about 40,000 years, which though long by human standards is a mere flash in the pan of geological time. The fossil record shows that at least some trends in physical evolution, once they become established, continue for long periods of time, even as measured by geological standards. In general, they continue because they are directly or indirectly influenced by natural selection. One such trend among humans and other primates as well has been (within broad limits) the evolution of a more complicated and more capable brain, which is reflected in solid bone by an increase in cranial capacity—that is, by an increase in the volume of the brain. Yet in the last 40,000 years (a period of time in which somewhat more than 2,500 human generations have come and gone), the volume of the human brain has not changed at all. It is probably safe to assume that science fiction writers are misleading us when they conjure up images of our descendants hundreds of centuries hence with overgrown brains encased in enormous globular skulls.

The fossil record also shows a slight tendency toward an increase in height in the more recent stages of human evolution, but this is probably due mostly to environmental rather than genetic changes. Minor changes in human teeth, with a slight tendency toward reduction in the number of wisdom teeth, may also have occurred in the past 30,000 years or so, though this is less well documented than the change in stature.

Other than that, as judged by the fossil record, the physical evolution of human beings appears to be at a standstill, or at least proceeding at a rate that is so slow as to be unnoticeable. Of course, only time will tell, but most of us will surely be long dead before any noticeable physical changes in the human species occur. Nonetheless, we have all experienced social and other cultural changes, some of which occur so fast that they bewilder us. Whatever its final basis, whether genetic or environmental, whether measured by IQ scores or by height, whether judged by cranial capacity or by other measures, the fact is that human beings of all races are wonderfully variable. And this enormous stockpile of human differences is our species' greatest asset for the future. Variability—whether physical, behavioral, biochemical, or otherwise—is what is important in the future evolution of any species, including our own.

Summary

Human behavior depends on the interactions of many genes and many environmental factors. Certain single-gene defects, such as the Lesch-Nyhan syndrome and PKU, can markedly influence behavior and frequently result in mental retardation. Most cases of mental retardation are of unknown cause and are presumed to be multifactorial.

Heritability is a statistical concept that estimates how much of the phenotypic variation in any population is due to genetic factors (nature) as opposed to environmental factors (nurture). Although it is impossible to accurately assess the heritability of any human trait, the study of one-egg twins raised apart provides a good estimate of the relative importance of genetic and environmental factors.

Human behavior is so varied and so changeable that it is usually impossible to sort out the genetic and environmental factors that influence it. Mental retardation, schizophrenia, and the ability to score high on IQ tests are behavioral traits known to be strongly influenced both by genetic and by environmental factors. Measurements of human behavior engender heated controversies, as evidenced by the dispute over the significance of racial differences in IQ scores and by the shutdown of a screening program for detecting XYY males.

Three types of natural selection are influencing the worldwide human population: prenatal selection, postnatal selection, and fecundity selection. The effects of natural selection on human body surfaces are no longer very important in human evolution, because people now adapt to the environment primarily by means of behavior, not body surfaces. Overall, natural selection acts as a stabilizing influence in human evolution, and our species has not undergone any noticeable physical change in the past 40,000 years at least. Yet social and other cultural changes in human behavior occur frequently and may take place very rapidly, as is well known. As long as people remain variable, they will continue to evolve.

Suggested Readings

"Intelligence and Race," by W. F. Bodmer and L. L. Cavalli-Sforza, *Scientific American*, October 1970, Offprint 1199. This review of the differences in IQ scores between black and white Americans concludes that the heritability of IQ scores cannot be accurately determined from the data presently available.

"Behavioral Implications of the Human XYY Genotype," by Ernest B. Hook. *Science*, vol. 179, 12 Jan. 1973. A review of some of the data on this complicated subject.

"Fragile Sites and X-Linked Retardation," by Frederick Hecht, Barbara K. Hecht, and Thomas W. Glover. *Hospital Practice*, November 1981. How physically weak portions of the X chromosome may contribute to many cases of mental retardation in males.

"Identical Twins Reared Apart," by Constance Holden. *Science*, vol. 207, 21 Mar. 1980. Provides some of the amazing details of the University of Minnesota twin study discussed in this book.

"XYY: Harvard Researcher Under Fire Stops Newborn Screening," by Barbara J. Cullington, *Science*, vol. 188, 27 June 1975. Discusses how social pressures stopped a screening program to detect sex chromosomes XYY in newborns.

"Criminality in XYY and XXY Men," by Herman A. Witkin et al. *Science*, vol. 193, 13 Aug. 1976. A detailed study that suggests that the elevated crime rate among XYY males is not related to aggression but may be related to low IQs.

"Genetic Influences in Criminal Convictions: Evidence from an Adoption Cohort," by A. M. Sarnoff et al. Science, vol. 224, 25 May 1984. Compares the incidence of criminal convictions among adopted children and their genetic or adopted parents in Denmark from 1927 to 1947.

Human Diversity, by Richard Lewontin. *Scientific American Books*, 1982. Provides excellent coverage of the genetics of IQ scores.

Some Genetics Problems Concerning Human Pedigrees

You may want to test your understanding of the pattterns of inheritance discussed in Chapters 1 and 2 by working the following problems. The answers are given at the end.

1. Huntington's chorea is a rare, fatal disease of the nervous system whose symptoms are not exhibited until middle age. It is inherited as an autosomal dominant trait. Suppose that an apparently normal man in his early twenties learns that his father has just been diagnosed as having Huntington's chorea.
    a. What are the chances that the son will eventually develop the disease?
    b. If the son never develops the disease, what are the chances that his offspring will have Huntington's chorea?

2. Maple syrup urine disease is a rare inborn error of metabolism that derives its name from the odor of the urine of affected persons. Those in whom the disease is untreated are severely mentally retarded, and they usually die as infants. The disease tends to recur in certain families, but the parents of affected children are always normal. Assuming that a single pair of alleles governs the occurrence of this disease, what does this suggest about the inheritance of maple syrup urine disease?

3. About 7 percent of whites in the United States cannot smell the odor of musk. If both parents cannot smell musk, all of their children will be unable to smell it. On the other hand, two parents who can smell musk generally have children who can also smell it; only a few of their offspring will be unable to smell musk. Assuming that a single pair of alleles governs this trait, what does this suggest about the inheritance of the ability to smell the odor of musk?[1]

4. Total color blindness is a rare condition that is inherited as an autosomal recessive trait. Affected persons see the world only in shades of gray and can see best in dim light or in the dark. A woman whose father is totally color-blind intends to marry a man whose mother was totally color-blind. What are the chances that they will produce affected offspring?

---

1. From *Genetics, Evolution and Man,* by W. F. Bodmer and L. L. Cavalli-Sforza. W. H. Freeman and Company, 1976.

5. As discussed in Chapter 1, albinism is inherited as an autosomal recessive trait. An albino man marries a normally pigmented woman and they have nine children, all normally pigmented. What are the genotypes of the parents and the children?

6. A normally pigmented man whose father was an albino marries an albino woman whose parents were both normally pigmented. They have three children, two normally pigmented and one albino. List the genotypes of all of these persons.

7. Achondroplastic dwarfism (see Figure 7–13) is an autosomal dominant trait. One would therefore expect two achondroplastic dwarfs to produce offspring in the ratio of about 75 percent dwarfs to 25 percent normal. Nonetheless, the observed ratio among the offspring of such matings is probably closer to 66 percent dwarfs and 33 percent normal. Can you account for the differences between the predicted and observed rates?

8. In humans, aniridia (absence of the iris, the colored part of the eye) can be a genetically determined cause of blindness that is transmitted as an autosomal dominant trait. Unilateral deafness (total deafness in one ear) can also be genetically determined and has an autosomal dominant pattern of inheritance. A man with genetically determined aniridia whose mother was not blind marries a woman with genetically determined unilateral deafness whose father has normal hearing. What proportion of their children would be expected to be *both* blind and deaf in one ear? (Hint: The probability that two independent events will occur together is the product of the individual probabilities.)

9. Suppose that a woman who has normal vision, but whose parents are both blind because of aniridia, consults a genetic counselor. She has unilateral deafness, as does her father. She wants to know what the chances are (assuming she marries a normal man) that her children will be either deaf in one ear or blind. What should the counselor tell her? (The patterns of inheritance of these two traits are given in Problem 8.)

10. A man who has sickle-cell trait marries a woman whose father had sickle-cell disease.
    a. Can they produce offspring who have neither sickle-cell disease nor sickle-cell trait?
    b. What proportion of their offspring will have sickle-cell disease?
    c. What proportion of their offspring will have sickle-cell trait?

11. A man of blood group O marries a woman of blood group A. The woman's father was of blood group O. What are the chances that their children will belong to blood group O?

12. In the following case of disputed paternity, which of the possible fathers can be excluded as the real father? The mother is of blood group B, the child is of blood group O, one possible father is of blood group A, and the other is of blood group AB.

13. Four babies were born in a hospital on a night in which an electrical blackout occurred. In the confusion that followed, their identification bracelets were mixed up. Conveniently, the babies are of four different blood groups: O, A, B, and AB. The four pairs of parents have the following blood groups: O and O, AB and O, A and B, B and B. Which baby belongs to which parents?[2]

14. A woman who has unusually short fingers marries a man who has fingers of normal length, and they have four children, two of each sex. One of their sons and one of their daughters have unusually short fingers.
    a. Draw the pedigree.
    b. If finger length is governed by a single pair of alleles, could the pedigree be explained by autosomal dominant inheritance?
    c. Could the pedigree be an example of autosomal recessive inheritance?
    d. Could the allele responsible for this trait be located on the X chromosome?

15. Assume that the following pedigree is for a trait that is rare in a particular population.

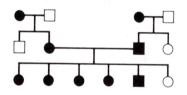

Indicate whether each of the following patterns of inheritance is consistent with or excluded by this pedigree.[3]
    a. Autosomal recessive.
    b. Autosomal dominant.
    c. X-linked recessive.
    d. X-linked dominant.
    e. Y-linkage.

16. Red–green color blindness is a relatively common condition that is inherited as an X-linked recessive. A normal woman whose father was red-green color-blind marries a man who has normal vision.
    a. What proportion of her sons would you expect to be red-green color-blind?
    b. If she married a man who was red-green color-blind, what proportion of their sons would you expect to have normal vision?

---

2. From *Heredity Evolution and Society*, by I. Michael Lerner and William Libby. W. H. Freeman and Company, 1976.

---

3. From *General Genetics*, by Adrian Srb, Ray D. Owen, and Robert Edgar. W. H. Freeman and Company, 1965.

c. If she married a man who was red-green color-blind, what proportion of their daughters would be carriers?

17. What pattern of inheritance best accounts for the following pedigree?

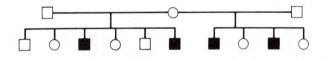

18. As discussed in Chapter 2, hemophilia is inherited as an X-linked recessive trait. An apparently normal woman whose father was a hemophiliac marries a normal man.
   a. What proportion of their sons will have hemophilia?
   b. What proportion of their daughters will have hemophilia?
   c. What proportion of their daughters will be carriers?

19. A man who is red-green color-blind (X-linked recessive) has four normal sons by his first wife and a color-blind daughter by his second wife. Both wives have normal vision. What is the genotype of each wife?

20. A short-fingered man (autosomal dominant) marries a woman who has normal hands and is red-green color-blind (X-linked recessive). List the possible phenotypes of their sons and daughters.

21. In the following pedigree, a dot represents the presence of an extra finger and a shaded area represents the occurrence of an eye disease.[4]

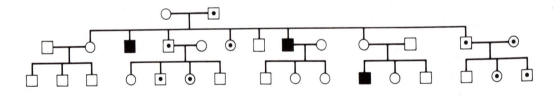

   a. What can you figure out about the inheritance of the extra finger?
   b. What two patterns of heredity might explain the inheritance of the eye disease?

22. Congenital deafness results when a person is homozygous for either or both of two recessive alleles, d and e. Both of the corresponding dominant alleles D and E are required for normal hearing, and the two pairs of alleles are inherited independently. A deaf man of genotype $ddEE$ marries a woman with normal hearing who is of genotype $DdEe$. What proportion of their offspring will be deaf?

4. From *An Introduction to Genetic Analysis,* 2d Ed., by David T. Suzuki, Anthony J. F. Griffiths, and Richard Lewontin. W. H. Freeman and Company, 1981.

23. Complete the following table.

Condition	Sex Chromosomes	Number of Barr Bodies
Normal woman		
Normal man		
	XO	
	XXY	
Down's syndrome		

24. Would you expect a person of sex chromosome constitution XXXX to have higher, lower, or approximately equal concentrations of G6PD compared to a person whose sex chromosomes are XXXY?

25. A woman who has Turner's syndrome is found to have hemophilia, yet neither of her parents has the disease. How is this possible?

26. In cats, the coat pattern known as tortoiseshell is made up of patches of black and orangish yellow. The alleles for black or orangish yellow coat color are on the X chromosome. In some regions of the coat, the X chromosome with the allele for black coat color is active; in other regions, the X chromosome with the allele for orangish yellow coat color is active. Hence, the tortoiseshell's patchwork coat is in good accord with Lyons' hypothesis, and one would therefore expect all tortoiseshell cats to be female. This is generally true, but on rare occasions male tortoiseshells are produced, and they are invariably sterile. Based on what you known about human sex determination and Lyons' hypothesis, can you explain the rare occurrence of tortoiseshell males?

27. How many Barr bodies would be found in a person who is a sexual mosaic with sex chromosomes XX/XXXY?

Answers

1. a. The chances are 50:50. Because the disease is rare, the affected father is most likely heterozygous (of genotype *Hh, H* for Huntington's chorea). All of his offspring therefore have a 50:50 chance of inheriting the dominant allele.
   b. Virtually zero, assuming that he marries an unaffected woman. Huntington's chorea can arise by spontaneous mutation, but this is a very rare event.

2. This pattern suggests that maple syrup urine disease is inherited as a recessive trait.

3. This pattern suggests that the ability to smell the odor of musk is dominant and that the inability to do so is recessive.

4. The chances are 1:4. The man and woman must both be heterozygous (of genotype Cc,) so these are the kinds of offspring they can produce:

	C	c
C	CC	Cc
c	Cc	cc

Thus, on the average, one-fourth of their offspring would be normal (CC), one-half would be carriers (Cc), and one-fourth would be totally color-blind (cc).

5. The man is an albino, so he must be homozygous recessive (of genotype aa). None of the nine children is an albino, but each must be a carrier (of genotype Aa). Because there are no albinos among the nine children, the mother is apparently of genotype AA.

6. The man must be of genotype Aa and his father of genotype aa. Similarly, the woman must be aa and her parents must both be Aa. Two of the children are of genotype Aa, and the other is aa.

7. As is true of nearly all autosomal dominant traits, individuals with achondroplastic dwarfism are heterozygous for the abnormal allele. That is, their genotype is Aa. When two Aa individuals produce offspring, about 25 percent of the offspring will be AA and will not survive to be born alive. The observed ratio among the offspring will therefore be about two dwarfs (Aa) to one normal (aa), or about 3:1.

8. Both parents are heterozygotes for one or the other autosomal dominant trait, so the probability that either of them will pass on the defective gene to their offspring is 50 percent. Therefore, the probability that a child will have both aniridia and unilateral deafness is $0.5 \times 0.5 = 0.25$ (25 percent).

9. The woman did not inherit the abnormal allele for aniridia from either parent, so she is normal with regard to it and there is no chance (except for the rare possibility of a mutation) that her children will be affected. Because she is affected by an autosomal dominant trait, there is a 50 percent chance that her children will be deaf in one ear.

10. a. Yes. Both parents must be heterozygous, so they can produce the following kinds of offspring:

	Hb^S	Hb^A
Hb^S	$Hb^S Hb^S$	$Hb^S Hb^A$
Hb^A	$Hb^S Hb^A$	$Hb^A Hb^A$

($Hb^A Hb^A$ = normal, $Hb^S Hb^A$ = sickle-cell trait, $Hb^S Hb^S$ = sickle-cell disease.)

    b. One-fourth (genotype $Hb^S Hb^S$).

    c. One-half (genotype $Hb^S Hb^A$).

11. The chances are $50:50$. The woman must be of genotype $I^A I^O$, so these parents can produce the following offspring:

	I^A	I^O
I^O	$I^A I^O$	$I^O I^O$
I^O	$I^A I^O$	$I^O I^O$

Thus, about one-half of their offspring will be of genotype $I^O I^O$ and will therefore belong to blood group O.

12. The man of blood group AB cannot be the real father. The mother is apparently of genotype $I^B I^O$. Her child is of group O and therefore must have inherited the O^O allele from each parent. The man of group AB can contribute only I^A or I^B and therefore cannot be the father. The man of group A could be of genotype $I^A I^O$, but he could also be of genotype $I^A I^A$. Thus, we can neither exclude the man of group A nor conclude that he is the real father.

13. a. Baby O could only belong to parents O and O, because the parents must be of genotype $I^O I^O$.
    b. Baby AB must belong to parents A and B becauuse only they could produce an offspring to genotype $I^A I^B$.
    c. Of the remaining two, baby A cannot belong to parents B and B, because they must be of genotype $I^B I^B$ or $I^B I^O$. Baby A must therefore belong to parents AB and O.
    d. Hence, baby B must belong to parents B and B.

14. a. The pedigree is this:

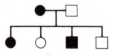

   b. Yes. In fact, this trait (officially known as *brachydactyly*) was the first autosomal dominant trait described among human families.
   c. Yes. But this is not likely to be an example of autosomal recessive inheritance because the father would have to be a carrier, which is unlikely.
   d. Yes. The trait cannot be inherited as an X-linked recessive because one of the sons does not manifest it. But the pedigree is consistent with X-linked dominant inheritance.

15. a. Autosomal recessive inheritance is excluded because the mating of two affected persons produced a normal daughter.
    b. The pedigree is consistent with autosomal dominant inheritance.
    c. X-linked recessive inheritance is excluded because an affected woman produced a normal son.
    d. X-linked dominant inheritance is also excluded because an affected man produced a normal daughter.
    e. Y linkage is excluded because of the presence of affected women.

16. a. One-half. The woman must be a carrier. All sons receive one or the other of their mother's X chromosomes, so the chances are 50:50 that any son would receive the chromosome with the abnormal allele.

    b. One-half. No sons receive their father's X chromosome, so the fact that the father is affected does not alter the chances that his sons will be red-green color-blind.

    c. One-half. All daughters receive their father's X chromosome and one or the other of their mother's X chromosomes. Thus, this marriage would produce color-blind and carrier daughters in roughly equal proportions.

17. X-linked recessive inheritance. The pedigree could also be explained by autosomal recessive inheritance, but that would require all three parents to be carriers of the same allele.

18. a. One half. The woman must be a carrier of hemophilia, and on the average half of her sons get the X chromosome that has the abnormal allele.

    b. None of them. An affected daughter could be produced only if the father were affected.

    c. One-half. On the average, half of the daughters receive the abnormal X chromosome from their mother and therefore are carriers.

19. The first wife is most likely normal, and the second wife must be a carrier of red-green color-blindness.

20. All sons will be red-green color-blind, and about half of them will have short fingers. All daughters will be carriers of red-green color blindness, and about half of them will have short fingers.

21. a. The extra finger is probably an example of autosomal dominant inheritance.

    b. The eye disease is probably inherited as an X-linked recessive trait, but it could be an example of autosomal recessive inheritance.

22. One-half. All of the man's sperm cells must contain the factors d and E. But the woman can produce four different kinds of eggs, namely, DE, De, dE, and de. So the following table can be constructed:

	dE	
DE	$DdEE$	(normal)
De	$DdEe$	(normal)
dE	$ddEE$	(deaf)
de	$ddEe$	(deaf)

23. The completed table is as follows:

Condition	Sex Chromosomes	Number of Barr Bodies
Normal woman	XX	1
Normal man	XY	0
Turner's syndrome	XO	0
Klinefelter's syndrome	XXY	1
Down's syndrome	XX or XY	1 or 0

24. The concentrations of G6PD in these two persons would be approximately equal. The XXXX individual has one active X chromosome and three Barr bodies, and the XXXY individual has one active X chromosome and two Barr bodies.

25. This could happen if the woman's mother were a carrier and nondisjunction resulted in the daughter's receiving no X chromosome from her father. If the daughter received only her mother's abnormal X chromosome, she would show symptoms of hemophilia.

26. The male tortoiseshell cats turn out to be the feline equivalents of human males with Klinefelter's syndrome—their sex chromosomes are XXY. Normal males (XY) are either black or orangish yellow. In the exceptional XXY males, one or the other X is active in various regions of the skin, and the coat is therefore a patchwork of black and orangish yellow—in other words, tortoiseshell.

27. One of two Barr bodies would be found, depending on which cell line is examined. Some of this person's cells are XX and therefore have one Barr body, whereas other cells are XXXY and therefore have two Barr bodies.

Appendix II

Some Simple Chemical Principles

This brief appendix is for people who have never studied chemistry or who have forgotten most of what they once knew about it. Although they are simplified, the definitions and other information given here should provide sufficient background for understanding Chapters 3 and 4.

All material objects are made up of *atoms*, and pure substances made up of only one kind of atom are known as *elements*. Elements can vary enormously in appearance and physical properties. For example, sodium is a soft, shiny metal that bursts into flame when dropped into warm water, bromine is a pungent, reddish-brown liquid, and gold is a lustrous metal that can be drawn out to form very thin threads. Of the 106 known elements, only a few are abundant in the human body. Over 90 percent of human body weight is accounted for by only four elements: oxygen, nitrogen, carbon, and hydrogen. Elements are assigned a one- or two-letter symbol based on their English or Latin names. The accompanying table lists the most abundant elements in the human body.

Although each element is associated with a unique kind of atom, all atoms have the same basic structure. A very dense central core, known as the *nucleus*, is surrounded by a much more extensive cloud of negatively charged particles known as *electrons*. Each kind of atom has a distinctive number of electrons in the cloud that surrounds the nucleus, and when atoms combine with each other to form molecules, they lose, gain, or share electrons. The resulting molecules are held together by forces known as *chemical bonds*, which link the atoms of the molecule together.

When elements combine to form molecules, the resulting pure substance, known as a *compound*, has markedly different properties than the elements that formed it. For example, the compound NaCl (sodium chloride, or table salt) results when an atom of sodium combines with an atom of the pungent, greenish-yellow gas, chlorine. Compounds are usually very stable substances because the atoms that combine to form the molecules of a given compound have more stable arrangements of their electrons when combined. In the example of sodium chloride, a more stable arrangement of electrons is formed by the loss of an electron from a sodium atom and the gain of an electron by a chlorine atom.

A *covalent bond* is a chemical bond that is formed when two or more atoms share one or more electrons. For example, the water molecule H_2O is the result of an atom of oxygen sharing electrons with each of the two hydrogen atoms. Using the appropriate symbols for the two elements and employing a dash to represent one pair of shared electrons, the water molecule can be expressed like this:

H—O—H

This kind of notation is known as a *structural formula*. Structural formulas indicate which atoms are attached and how many electrons they share. (Two pairs of shared electrons are indicated by two dashes ($=$) three pairs of shared electrons by three dashes ($\equiv$). For example, the amino acid cysteine, one of the 20 basic building blocks of proteins, has the following structural formula:

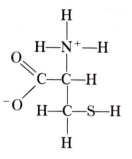

As you can see, a central carbon atom shares a pair of electrons with four other atoms—two carbons, one nitrogen, and one hydrogen. The lowermost carbon atom forms single covalent bonds (sharing one pair of electrons) with two hydrogen atoms and one sulfur atom, and so on. The only double covalent bond (formed by the sharing of two pairs of electrons) is between the carbon atom on the left and one of the oxygen atoms, and the $+$ and $-$ signs indicate that an excess positive or negative electrical charge is present at a given point within the molecule. For convenience, covalent bonds involving hydrogen atoms are usually not written out, so that the commonly employed structural formula for cysteine is as follows:

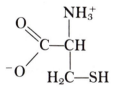

Structural formulas, however, tell us nothing about the exact shapes of the molecules they represent. For that purpose, *space-filling models* are used. Here are the space-filling models for water and cysteine:

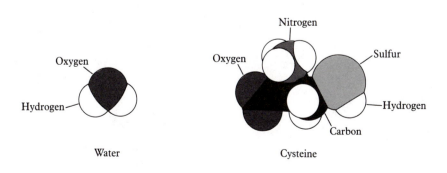

In Chapters 3 and 4, structural formulas and space-filling models are both used to explain the structure and function of the human genetic program.

Most of the more complex molecules found in living things are *polymers*, which means that they are made up of long chains of building blocks that are themselves molecules of a particular kind. Thus, DNA (a nucleic acid) is made up of nucleotide building blocks (sugar + phosphate + nitrogen-containing base), and proteins consist of long chains of amino acids joined by chemical bonds. The building blocks (*monomers*) always join in the same way. For example, the amino acids of proteins are joined by *peptide linkages*, which are special kinds of covalent bonds that form between a carbon atom in one amino acid and a nitrogen atom in another. When the amino acids glycine and alanine are joined, the formation of the peptide linkage can be represented as follows:

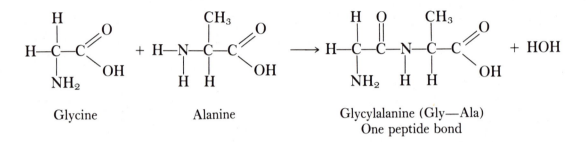

Glycine   Alanine   Glycylalanine (Gly—Ala)
                                      One peptide bond

All of the peptide linkages in a given protein are formed in the same way, and a molecule of water is produced each time one is formed.

In spite of their enormous variety, practically all of the chemical reactions that occur in the human body, which are collectively referred to as *metabolism*, have two things in common. First, they depend on specific enzymes, without which they would not take place to any appreciable degree. Second, the enzyme-dependent chemical reactions of human metabolism are interconnected by intricate *metabolic pathways*, in which a compound formed by one reaction serves as one of the starting substances of the next reaction in line. Enzymes and metabolic pathways are discussed in Chapter 3.

Index